DISCRETE STRUCTURES
OF
COMPUTER SCIENCE

LEON S. LEVY
Bell Laboratories
Whippany, N.J.

JOHN WILEY & SONS
NEW YORK · CHICHESTER · BRISBANE · TORONTO

For Millie

QA
248
.L43

Library of Congress Cataloging in Publication Data:

Levy, Leon S
    Discrete structures of computer science.

    Includes index.
    1. Set theory.   2. Functions.   3. Graph theory.
4. Algebra.   5. Electronic digital computers—Programming.   I. Title.
QA248.L43   519.4   79-11218
ISBN 0-471-03208-5

Printed in the United States of America

10 9 8 7 6 5 4 3 2 1

PREFACE

The notes on which this textbook is based have been used for a second-year, one-semester course in discrete structures of computer science at the University of Delaware. The requirement for such a course evolved from a continuing reevaluation of curricula in computer science; I then concluded that a sound appreciation of the underlying foundations is essential to good programming practice.

The computer science curriculum at the University of Delaware has been stabilized around the following nucleus of courses that is necessary to obtain an undergraduate degree. First-year students learn a programming language and get some problem-solving experience. Second-year students study discrete structures and basic mathematical logic. At the junior level, several courses are available, including machine structures, automata and formal languages, and programming languages and systems programming. The senior year is devoted to more advanced topics (for example, compiler writing) and to programming projects.

This curriculum sketch is important to an understanding of the role of the discrete structures course. In each of the junior-level courses, the material has been the basis for a concise and accurate description of material. The increased "mathematical maturity" also gives students a deeper appreciation of concepts so that senior-year students are able to utilize the current literature.

The rationale for this curriculum is that computer science is partially a science of clear and concise description of computable, discrete sets. Since modern programming languages are not always appropriate for descriptions of algorithms, part of the computer scientists' training involves learning to formulate an algorithm in some mathematical form. I hope to introduce a foundation here.

My choice of topics has been prescribed somewhat by the intended level of the course and its place in the curriculum. My treatment of graph theory deals almost exclusively with directed graphs. Further study of graph theory can be pursued in courses on graph theory, algorithms, or combinatorics. Similarly, topics in combinatorics are likely to form a separate course; therefore, most counting exercises are kept implicit by including them in the problems. An exception is the principle of inclusion and exclusion, which seems to fit naturally into the discussion of set theory.

I have used a layered approach in teaching. Each chapter reinforces the preceding chapters. For example, in discussing formal systems in Chapter 4, I relate them to algebras, shown in Chapter 3; trees, presented in Chapter 5, are examined algebraically, related to formal systems and automata. At each stage, students should have added depth and increased knowledge. When the end of the text is reached, students should be ready to read original literature and use the introductory sections of advanced textbooks for notational variations only because of familiarity with the fundamental concepts.

There is more than enough material here for most one-semester, sophomore-level courses in computer science. In general, I have been able to cover Chapters 0 to 4 quite completely and to add selected material from Chapter 5 in a three-credit, one-semester course. Students who have covered this material are then capable of reading Chapters 5 and 6 on their own or as assigned material in junior- and senior-level courses.

Learning the material well requires the completion of many exercises by students. These exercises have been graded by adding an E for a relatively straightforward exercise, P for a somewhat more difficult problem, and an asterisk (*) for the few hard problems.

Several programming problems are included in each chapter. I found that programming is not especially helpful in learning this material, but the students must understand how formal computer science is useful in developing the abstract versions of algorithms.

Many people helped me in the preparation of this book. Professors Peter Buneman, John W. Carr, III, A. Toni Cohen, Martin Freeman, Saul Gorn, Amir Pnueli, Walter Jacobs, and Giora Slutzki all read the manuscript and offered helpful suggestions. Professors Carr and Gorn also taught me computer science. Professor Aravind Joshi has a special interest in Chapter 5, because we jointly studied tree automata. Lillian Cassel taught from earlier versions of the text and offered helpful suggestions. Finally, I thank the many students who were often both the forge and the anvil that shaped this text; Bob Melville and my son, Jordan, deserve special thanks.

PREFACE

The final manuscript was excellently typed by Alison Chandler; Vicki Calvert, Beverly Crowl, Debbie Young, and Karen Tanner all typed parts of the preliminary manuscript.

Leon S. Levy

CONTENTS

CHAPTER 5

CHAPTER 6

INDEX

an essay on discrete structures

SUMMARY

We begin our study by discussing briefly what computer science is, how formal methods are related to computer science, and some of the effects of using formal methods in computer science. Two examples will be used in this chapter to clarify the discussion.

Our chapters start at Chapter 0 because we are using the natural numbers for counting.

Definition
The natural numbers are the integers that are greater than or equal to 0: 0, 1, 2, ...

A. A VIEW OF COMPUTER SCIENCE

Computer science organizes information about the way we calculate. A simple view of calculation would restrict it to talking about numbers. A more general view includes calculating with symbols. Even more generally, we could allow procedures such as those described in Figure 0-1 to be studied.

Having perhaps tentatively decided to include such procedures, have we gone too far? Is the block "GET GAS" well enough defined to be included in our study? One may argue the merits and disadvantages as follows.

No	*Yes*
It is too vague:	If necessary, additional
What kind of gas?	detail can be added
How much?	to the flowchart.
What gas station?	
⋮	

Even if we restrict ourselves to symbolic processes where the computer is circumscribed only to read, print, and calculate using symbols from some fixed character set, we will be faced with many situations where the boundaries of computer science are difficult to draw. Can reading a novel and writing a report about it or writing a poem be usefully studied in computer science?

We cannot always answer the question as to what can

Figure 0-1.   Drive to work.

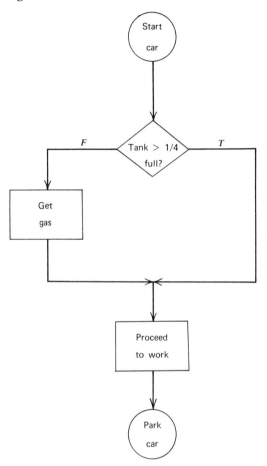

usefully be studied in computer science, but we can and will limit our study to areas that are well-defined and generally accepted to be within the purview of computer science.

We say that computer science is, for our purposes, *the study of the algorithmic method of calculation*, where an algorithm is, informally, a computational recipe. In Chapter 4, a more technical formulation of this idea is presented.

A more general view of computer science is that it is the study of programming. Within this comprehensive description, the various software and hardware tools would necessarily be studied. There has, indeed, been a considerable amount of programming done both to synthesize and to understand stories and poems. If we neglect these here, it is *not* because we exclude from computer science the programs that include heuristics, but because formal methods, such as the ones in this book, deal with simplifications of actual objects, and we do not know the necessary idealizations for studying many of the items that are of interest to computer scientists.

B. ON FORMAL METHODS

Formal methods require stylized descriptions that are often only an approximation of the actual objects to which they are supposed to correspond. Nevertheless, formal methods are useful because they provide convenient abstractions and language for organizing our relationship to the world.

Two examples of formal methods are given next to illustrate different aspects of formalism. In the first method, mathematical induction, we illustrate a method that (together with its general-izations, which are presented in Chapters 4, 5, and 6) finds many applications in computer science. The second method, analogy, also has many applications; and it is specially illustrated in algebraic methods through the concept of a *homomorphism*, which is defined and discussed in Chapter 3. (Intuitively, a homomorphism is like a description of the relation between an object and its shadow, but its technical exposition must be deferred.)

Induction

Consider the program that is flowcharted in Figure 0-2. Suppose 5 is read in as the value of *n*. Let us trace the computation and see what is computed.

	n	x	y
Leaving block 1	5	1	0
First time leaving block 2	4	3	1
Second time leaving block 2	3	5	4
Third time leaving block 2	2	7	9
Fourth time leaving block 2	1	9	16
Fifth time leaving block 2	0	11	25

Figure 0-2. Computation of n_0^2.

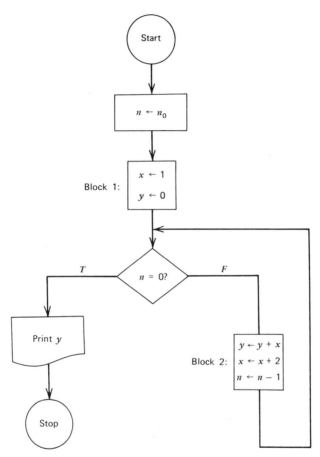

What can be said about this flowchart and about a program derived from it? The flowchart includes the symbolic numerals 0, 1, 2, the operation symbols +, −, an = symbol, and an assignment arrow. We understand that in the flowchart the numerals represent numbers, and certain properties of these numbers are assumed; for example, $1 − 1 = 0$ is always true. As long as the program corresponding to our flowchart preserves the assumed properties, we expect that the results of the program will correspond to the computation specified by the flowchart. In particular, if very large values of n are used that are larger than the ones that our programming language is prepared to handle exactly, we would anticipate that the program and the flowchart would not agree.

Assuming that for n an integer, $n \leq 1,000,000$, the program and flowchart are equivalent, we expect that if n_0 is nonnegative where n_0 is the value of n read in by the program, then:

The loop that includes block 2 *is traversed exactly* n_0 *times;*
The value of x, when the true exit is taken, is $2n_0 + 1$;
The value of y, each time block 2 is left, is a perfect square ≥ 0;
$[(x - 1)/2] + n = n_0$, whenever the test, "$n = 0$?", is entered.

Moreover, all of these assertions can be inferred from the flowchart for whatever maximum value of integer can be accommodated by the program. How? In the case of Figure 0-2 one is tempted to say "by inspection." However, we should know in more detail the reasoning behind the "obvious" assertions so that similar reasoning may be applied to less transparent situations. Also, we should be able to generalize the ideas. Finally, we should be able to justify our insights better to ourselves and to others who may not see that what we claim is obvious.

Each of the claims about the algorithm of Figure 0-2 is potentially a claim about an infinite number of cases and requires a method for justifying such claims. The method is known as *mathematical induction*, which is suggested by the fact that, from a finite number of observations, inferences about an infinite number of cases are being drawn. The justification for the method is called the *principle of mathematical induction. Let P(x) be a statement about x. If P(0) is true, and if from P(n) we can infer P(n + 1), then from both of these facts we may conclude that P(x) is true of every natural number x.*

In order to use the method of mathematical induction, two things must be demonstrated. The first is that P is true about 0; this is called the *basis*. Second, that assuming that P is true about n, it follows that P is true about $n + 1$; this is called *induction*. We demonstrate the method of mathematical induction for some of the assertions about Figure 0-2.

The loop that includes block 2 *is traversed exactly* n_0 *times.*
We use induction to prove a slightly different statement that directly implies that the loop is traversed n_0 times: if n has a

value p on reaching the test, exactly p traversals of the loop will follow.

BASIS. If n is 0, the *true* exit is taken from the test immediately and the loop, including block 2, is traversed 0 times.

Induction. Assume that if the initial value of n is p, we traverse the loop p times. Then, if n is $p+1$ initially, the loop is traversed once and n is reduced from $p+1$ to p. By the inductive assumption, p additional traversals occur.

The value of x, when the true exit is taken, is $2n_0 + 1$.
Again we prove the following slightly different statement.

After k traversals of the loop, the value of x is $2k + 1$.

BASIS. In block 1, x is assigned a value of 1. If k is 0, a *true* exit occurs immediately, and

$$x = 1$$
$$= 2 \cdot 0 + 1$$
$$= 2k + 1 \qquad \text{since} \qquad k = 0$$

Induction. Each time the loop is traversed, x is increased by 2. Let $x(k)$ denote the value of x after k traversals. Assuming that $x(k) = 2k + 1$, the value of x after $n + 1$ traversals is $x(k + 1)$. Since we add 2 to x during each traversal,

$$x(k + 1) = x(k) + 2$$
$$= (2k + 1) + 2 \qquad \text{by the inductive assumption}$$
$$= 2(k + 1) + 1 \qquad \text{rearranging terms}$$

The value of y, each time block 2 is left, is a perfect square.
In this case, we must strengthen the assertion to $y = k^2$, where k is the number of times the loop has been traversed, and we use the fact that the kth time we enter the test block y has the same value it has on the kth exit from block 2.

BASIS. If $k = 0$, we have not been around the loop and $y = 0$, so $y = 0^2$.

Induction. Let $y(k)$ denote the value of y at the kth step. Then the inductive assumption is $y(k) = k^2$ and, using this assumption, we must show

$$y(k + 1) = (k + 1)^2$$

We know that

$$y(k + 1) = y(k) + x(k) \qquad \text{from the flowchart} \qquad (0\text{-}1)$$

$$x(k) = 2k + 1 \qquad \text{which was first proved} \qquad (0\text{-}2)$$

Using the inductive assumption that $y(k) = k^2$ and substituting Equation 0-2 for $x(k)$ in Equation 0-1, we have

$$y(k + 1) = k^2 + 2k + 1$$
$$= (k + 1)^2 \qquad \text{(factoring)}$$

The proof of the last assertion, that $[(x - 1)/2] + n = n_0$, is left for Problem 1.

Mathematical induction has been presented here and will be discussed again in Chapter 4. Our purpose here is to suggest the types of demands that we can make on our formalism. Consider the flowchart shown in Figure 0-3.

The proof that this flowchart computes squares and cubes is similar to the proofs about Figure 0-2. In fact, we can clearly apply the mathematical induction to higher-degree polynomials.

The flowcharts in Figures 0-2 and 0-3 have had the same control structures, and we should be able to vary the method for different control structures. We can assert that in each case the computation ends, since we can compute the number of traversals of the loop for each value of the input variable, while a more useful inductive argument would not require such an explicit form.

The inductive arguments have been specialized to computation over the natural numbers but, when we allow symbolic computation, more general principles of induction will be required. Indeed, a strongly recurring pattern in formal methods is taking a rule that is known to work in one case and generalizing the rule to apply to a much larger class.

Figure 0-3.   Computation of n_0^3.

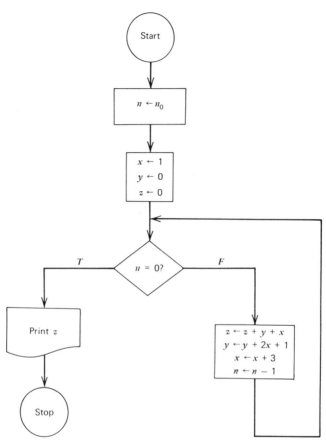

C. ANALOGY

An analogy is a correspondence between dissimilar objects. Such a correspondence can be quite useful; if two problems are analogous and we know how to solve one of them, we might be able to use the analogy to adapt the solution to the other problem. For example, you can easily adapt the rules for performing arithmetic in the decimal system to other analogous number systems over a different base. We will show a slightly more complicated example of an analogy.

The Gray code is a number representation system using k binary digits to represent numbers from 0 to $2^k - 1$. The one-digit Gray code represents the number 0 as the numeral 0 and the number 1 as the numeral 1. The Gray code over $k+1$ digits may be derived from the Gray code for k digits as follows. Let $G_k(x)$ denote the k digit representation of x in Gray code; then

$$G_{k+1}(x) = 0\ G_k(x) \qquad\qquad 0 \leqq x < 2^k$$
$$\phantom{G_{k+1}(x) =} 1\ G_k(2^{k+1}-1-x) \qquad 2^k \leqq x < 2^{k+1}$$

Example

The two-digit Gray codes are obtained from these formulas as:

$$G_2(0) = 0\ G_1(0) = 00$$
$$G_2(1) = 0\ G_1(1) = 01$$
$$G_2(2) = 1\ G_1(2^2-1-2) = 1\ G_1(1) = 11$$
$$G_2(3) = 1\ G_1(2^2-1-3) = 1\ G_1(0) = 10$$

Example

The five-digit Gray code for 12 is 01010 or, in our notation, $G_5(12) = 01010$. The six-digit Gray codes for 12 and 51 are, therefore, given by

$$G_6(12) = 0\ G_5(12) = 001010$$
$$G_6(51) = 1\ G_5(2^6-1-12) = 1\ G_5(12) = 101010$$

A convenient way to construct the sequence of $k+1$-digit Gray codes from the sequence of k-digit Gray codes is to copy the sequence, prefixing a 0 to each Gray code, and then copy the sequence again, this time in reverse order, prefixing a 1 to each Gray code. The construction is shown in Figure 0-4, where the construction of the four-digit Gray code sequence is derived from the three-digit Gray code sequence.

A useful property of the Gray code is that its representations of successive integers differ by exactly one binary digit. [As an example, referring to Figure 0-4, $G_4(6) = 0101$, $G_4(7) = 0100$, and $G_4(8) = 1100$. $G_4(6)$ and $G_4(7)$ differ in the rightmost digit of the representation, while $G_4(7)$ and $G_4(8)$ differ in the leftmost digit of the representation.] This property is useful in a number of applications, and different methods for generating the Gray code are, therefore, of interest.

One way to generate the Gray code sequence is to generate the binary number sequence and translate binary numbers into Gray code numbers. When the binary numbers are needed for

Figure 0-4.

0	0	0	0
1	0	0	1
2	0	1	1
3	0	1	0
4	1	1	0
5	1	1	1
6	1	0	1
7	1	0	0

Three-digit Gray code

1	0	0
1	0	1
1	1	1
1	1	0
0	1	0
0	1	1
0	0	1
0	0	0

Three-digit Gray
code in
reverse order

0	0	0	0	0
1	0	0	0	1
2	0	0	1	1
3	0	0	1	0
4	0	1	1	0
5	0	1	1	1
6	0	1	0	1
7	0	1	0	0
8	1	1	0	0
9	1	1	0	1
10	1	1	1	1
11	1	1	1	0
12	1	0	1	0
13	1	0	1	1
14	1	0	0	1
15	1	0	0	0

0's preceding three-digit Gray code
sequence

1's preceding three-digit Gray code
sequence in reverse order

Four-digit Gray code

other purposes this method may be useful. Let $\oplus$ denote the binary operation whose operation table is as follows.

$\oplus$	0	1
0	0	1
1	1	0

The rule for translating a binary number, $B_n = b_1b_2 \ldots b_n$, to a Gray code number, $G_n = g_1g_2 \ldots g_n$, is:

$$g_1 = b_1$$

and

$$g_k = b_{k-1} \oplus b_k \qquad 1 < k \leqq n$$

Example

The six-digit binary representation of 27 is 011011, where $b_1 = b_4 = 0$ and $b_2 = b_3 = b_5 = b_6 = 1$.

The Gray code is given by

$$g_1 = b_1 = 0$$
$$g_2 = b_1 \oplus b_2 = 0 \oplus 1 = 1$$
$$g_3 = b_2 \oplus b_3 = 1 \oplus 1 = 0$$
$$g_4 = b_3 \oplus b_4 = 1 \oplus 0 = 1$$
$$g_5 = b_4 \oplus b_5 = 0 \oplus 1 = 1$$
$$g_6 = b_5 \oplus b_6 = 1 \oplus 1 = 0$$

Thus the six-digit Gray code representation of 27, $G_6(27)$, is 010110. #

Chinese Rings Puzzle

This method of obtaining the Gray code representation requires the binary representation, which may not be available. So a more direct method of generating the Gray code sequence is desirable. We develop such a method from an analogous counting mechanism, known as the Chinese ring puzzle, which is shown in Figure 0-5. The puzzle is usually presented with all the rings on the horizontal bar (Figure 0-5a), and the object of the puzzle is to remove all the rings from the bar to separate the horizontal bar from the rings assembly (Figure 0-5b). Analysis of the ring puzzle mechanism shows that in any configuration, at most two operations are possible.

1.   One may at any time remove the first ring if it is on the horizontal bar, or one may move the ring onto the horizontal bar if it is not on it.

2.   If some ring other than the first one is the leftmost ring on the horizontal bar (oriented as shown in Figure 0-5a), the ring immediately to its right may be removed if it is on the bar, or moved onto the bar if it is not.

The sequence of configurations needed to solve the ring puzzle may be represented schematically as a sequence of arrays

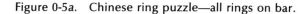

Figure 0-5a.   Chinese ring puzzle—all rings on bar.

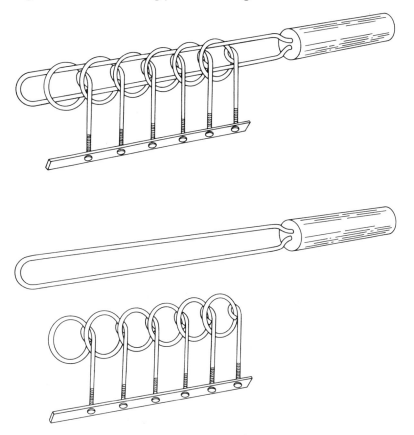

Figure 0-5b.   Chinese ring puzzle—all rings off bar.

of binary digits, in which each binary array represents a configuration by digits $r_1 r_2 \ldots r_n$, where r_i stands for the ith ring, and
$r_i = 1$ if ring i is on the horizontal bar and $r_i = 0$ if ring i is off the
horizontal bar.

The fact that the sequence of ring configurations changes by
a single binary digit at each step suggests that a ring puzzle
solution may provide a means for generating a Gray code
sequence. Indeed, the optimal configuration sequence for the
ring puzzle with four rings is given in Figure 0-6, and is seen, by
comparing it with Figure 0-4, to be a Gray code sequence in
reverse (but not all of the Gray code sequence; see Problem 5).

Figure 0-6.   Sequence of steps in solution of ring puzzle.

Starting configuration	1	1	1	1	Remove ring 2
	1	1	0	1	Remove ring 1
	1	1	0	0	Remove ring 4
	0	1	0	0	Add ring 1
	0	1	0	1	Add ring 2
	0	1	1	1	Remove ring 1
	0	1	1	0	Remove ring 3
	0	0	1	0	Add ring 1
	0	0	1	1	Remove ring 2
	0	0	0	1	Remove ring 1
	0	0	0	0	

The interesting fact latent in the ring puzzle is that in any configuration there are at most two moves possible, and that one of these moves just restores the preceding configuration. Thus a simple rule for the rings puzzle is to start by removing ring 2 and proceed by making any legitimate move, but never move the same ring twice in succession. By analogy, the Gray code may be generated by the following rule. *Start at* 000 . . . 0. *At any step, either change the rightmost bit, or change the bit to the left of the rightmost* 1. *Never change the same position twice in succession. At the first step, change the rightmost bit.*

The description of the Gray code sequence construction given initially was a recursive program; the Gray code sequence for $k+1$ digits was described in terms of the Gray code sequence for k digits. The subsequent analogy of the ring puzzle develops the Gray code sequence in a nonrecursive program. This suggests that other recursive programs may have similar nonrecursive equivalents. We illustrate this by developing an analogy between the Gray code sequences and the Towers of Hanoi puzzle, which is often used as a standard illustration of a beginner's recursive program.

Towers of Hanoi Puzzle (see Figure 0-7)

The puzzle consists of a set of three pegs and eight disks of increasing size, with the disks all initially on peg 1 in strictly increasing size, the smallest disk on top. By a sequence of moves transfer all the disks to peg 2, where a move consists of

Figure 0-7.   Towers of Hanoi puzzle.

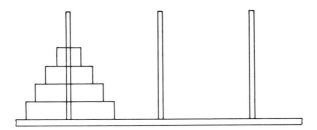

taking the topmost disk off a peg and placing it on another peg whose topmost disk is larger than the disk being moved.

The recursive program for the Towers of Hanoi is based on the fact that a move of the kth disk from peg i to peg j requires, as a precondition, that all disks smaller than the kth disk be on the remaining peg. The program is then based on the recursive algorithm.

Move disks 1 to k from peg i to peg j:
  if $k = 1$ <u>then</u> transfer disk 1 from peg i to peg j
        <u>else</u> <u>do</u> let $m = 6 - i - j$;
            move disks 1 to $k - 1$ from peg i to peg m;
            transfer disk k from peg i to peg j;
            move disks 1 to $k - 1$ from peg m to peg j
        <u>end do</u>

Note that in this program we have used the fact that if i and j are two different integers between 1 and 3, $6 - i - j$ is the third integer.

We now observe the following connection between the solution to the Towers of Hanoi puzzle and the Gray code sequence. If in the pth step of the solution to the Towers of Hanoi, the kth disk is moved, then in the pth step of the Gray code sequence the kth digit will be changed. Hence a solution to the Towers of Hanoi can be used in a generator of the Gray code sequence. Based on this observation, the following recursive Gray code generator program results; it lists the sequence of positions in which successive changes are made to generate the Gray code sequence.

Generate k-digit Gray code sequence:

<u>if</u> $k = 1$ <u>then</u> write "1"
<u>else do</u> Generate $k - 1$-digit Gray code sequence; write "k";
        Generate $k - 1$-digit Gray code sequence
        <u>end do</u>

The outputs of this program are shown in Figure 0-8 with the corresponding Gray codes for $k = 2, 3, 4$.

The analogy between the Gray code and the Towers of Hanoi can also be used to develop nonrecursive solutions to the Towers of Hanoi puzzle. We can see that given a Gray code sequence, we have the sequence of disks moved in the solution to the Towers of Hanoi puzzle; that is, disk i is moved when a change in bit i occurs. The remaining information needed to complete the specification of the move in the Towers of Hanoi puzzle solution is easy (see Problem 8).

Output List	Gray Code	Output List	Gray Code	Output List	Gray Code
	00		000		0000
1	01	1	001	1	0001
2	11	2	011	2	0011
1	10	1	010	1	0010
	$k = 2$	3	110	3	0110
		1	111	1	0111
		2	101	2	0101
		1	100	1	0100
			$k = 3$	4	1100
				1	1101
				2	1111
				1	1110
				3	1010
				1	1011
				2	1001
				1	1000
					$k = 4$

Figure 0-8. Sequence of outputs from recursive position generator and corresponding Gray codes.

Since we know that a nonrecursive solution exists for the Towers of Hanoi puzzle, we are motivated to find a direct solution that does not require an explicit Gray code generator. The nonrecursive Gray code generator previously developed was based on the observation that one should never change the same bit twice in succession. Using this fact in the Towers of Hanoi solution, which now means not moving the same disk twice in succession, we have the following program.

<u>do forever</u> move disk 1 to the next peg;
<u>comment</u>:   disk 1 moves cyclically on the pegs—from peg 1
            to peg 2 to peg 3 to peg 1, etc.   <u>end comment</u>
examine the two pegs other than the one holding disk 1;
<u>if</u> both are empty <u>then</u> halt;
<u>if</u> only one is nonempty move its topmost disk onto the other;
<u>else</u> compare the topmost disks on the two pegs and move
      the smaller onto the larger
<u>end do</u>

This program may be incorporated into a nonrecursive Gray code generator (see Problem 9).

Pursuing the analogy from a slightly different point of view, we note that all disks start on peg 1; knowing how many times a disk has been moved, we can determine its position uniquely. If the disk is an odd-numbered disk (counting from the top), count cyclically 1-2-3-1-2-3- . . . for the number of moves the disk has made to determine the peg on which it rests. If the disk is an even-numbered disk, count cyclically 1-3-2-1-3-2- . . . . It follows that the number of times that the disk has been moved uniquely determines its position. And, by analogy, the Gray code bit i is 1 if and only if disk i has been moved an odd number of times. We use this analogy to determine the configurations of both the Towers of Hanoi after the kth move and the corresponding Gray code value. Noting that the number of moves made by disk i is either the combined sum of the numbers of moves made by all disks larger than i or one more than that combined sum, we have the following number system for positive integers, which we call the *TH* number system, in which each number is represented as a list.

Let $TH(x)$ be the TH representation of x, and let $x_0 = \left\lfloor \dfrac{x+1}{2} \right\rfloor$. [$\lfloor z \rfloor$ is the largest integer less than or equal to z; its programming version is often denoted *entier* (z).] We define $TH(x)$ as follows.

$$TH(1) = 1$$

$$TH(x) = x_0, \; TH(x - x_0) \qquad \text{if} \qquad x > 1$$

The definition of $TH(x)$ is recursive, but will always contain a finite number of integers for any fixed x.

Example

(a) $TH(23) = 12, \; TH(11)$
(b) $TH(11) = 6, \; TH(5)$
(c) $TH(5) = 3, \; TH(2)$
(d) $TH(2) = 1, \; TH(1)$
(e) $TH(1) = 1$
(f) $TH(2) = 1, 1$ substituting (e) in (d)
(g) $TH(5) = 3, 1, 1$ substituting (f) in (c)
(h) $TH(11) = 6, 3, 1, 1$ substituting (h) in (b)
(i) $TH(23) = 12, 6, 3, 1, 1$ substituting (i) in (a) #

The number system $TH(x)$ has the property that the sequence representing any number is unique and, moreover, the kth integer in the sequence representing x is either the sum of all remaining integers in the sequence or one more than that sum. Thus, $TH(x)$ can be interpreted as a list of the number of moves made by successive disks in the Towers of Hanoi puzzle.

The Gray code is derivable from the Towers of Hanoi solution by defining $G(x)$, obtained from $TH(x)$ by replacing even integers in the sequence by 0 and replacing odd integers in the sequence by 1 (the leftmost digit of $G(x)$ is the least significant digit of the Gray code) and adding zeros if necessary for the most significant bits.

Example

$$TH(23) = 12, 6, 3, 1, 1$$
$$\updownarrow \; \updownarrow \; \updownarrow \; \updownarrow \; \updownarrow$$
$$G(23) = \; 0, 0, 1, 1, 1$$

and so the five-digit Gray code for 23 is 11100. #

D. ABSTRACTION

Applying formal methods in computer science requires a suitable set of abstractions. The ways in which we think about and describe our world are determined largely by the set of concepts into which we classify items and events. Thus, for example, it is known that the range of colors used to describe objects depends on the culture. Similarly, family relationships are defined in incomparable forms in different societies.

It is quite impractical to try to operate without idealized and simplified concepts. And the influence of the concepts and the operations to which they are subject on the conclusions that may be drawn motivates the development of specific meanings in the usage of terms. Scientific definitions of force, energy, pressure, and other physical quantities are familiar examples of well-defined terminology. Conversely, in the political sphere confusion often arises from lack of agreement or consistency in the use of basic words such as freedom, democracy, and totalitarianism.

In computer science, we also have the need for precise terminology with agreed definitions. The set of abstractions is useful both in discussing the objects whose properties are being transformed by algorithms and the algorithms themselves. Thus, a directed graph (defined in Chapter 2) can be an abstraction of a problem (e.g., finding a path through a maze), and a directed graph can also be an abstraction of an algorithm (for solving the maze).

The difficulty of defining concepts appropriately can be seen by considering our (as yet unformalized) notion of an algorithm. It often occurs that the same abstract algorithm underlies a variety of embodiments of the algorithm in different programs. In such cases, the theory tries to clarify how the different algorithms are related.

Example

The Euclidean algorithm for computing the greatest common divisor, gcd, of a pair of integers is often used as the paradigm of an algorithm. However, there is not one accepted form of this algorithm, and one often finds quite different formulations of this algorithm, such as the following ones.

```
gcd₁(x,y);
begin while x ≠ y do
        begin u : = min(x,y); v : = max(x,y); x : = u; y : = v − u
            end
        gcd₁ : = x
    end


gcd₂(x,y);
begin if x = y then gcd₂ : = x
        else begin          u := max(x,y);          v := min(x,y);
        gcd₂ : = gcd₂(u, v − u)
                end
    end
end
```

These two versions of the same abstract algorithm are similar, but they are clearly not identical. gcd₁ can be programmed directly in a language that does not allow recursion as a control construct, while gcd₂ cannot.   #

REVIEW

Computer science, in general, is defined as the study of computation. In this more formal framework, the scope of computer science studied is limited to algorithmic methods, but the precise definition of an algorithm is itself deferred.

Formal methods are illustrated by two general techniques: induction and analogy. Induction allows the proof of an infinite number of cases by reasoning about a finite number of those cases. An example is given of applying induction to prove the correct behavior of a flowchart program. Analogy allows the solution of one problem to be modified and applied to a second problem once the similarity of the two problems is grasped. Analogy is illustrated by showing the similarity between a program to generate Gray code and a program to solve the Towers of Hanoi puzzle.

Finally, the role of abstraction in computer science is compared to the role of abstraction in other sciences and, as the Euclidean algorithm demonstrates, the definition of concepts, such as algorithm, is often problematic.

PROBLEMS

E 1. Referring to the program in Figure 0-2, show that each time the program execution leaves block 2, the equation $[(x - 1)/2] + N = N_0$ is true.

E 2. Show, by mathematical induction, that the flowchart in Figure 0-3 computes z as the cube of n_0.

P 3. The binary operation $\oplus$ is used in the text to describe the Gray code in terms of the binary code. Show that if $x = y \oplus Z$, $y = x \oplus Z$. Using this fact, develop a set of equations for describing binary code in terms of the Gray code.

E 4. Verify that the Gray code construction given in Figure 0-4 satisfies the recurrence formula for Gray codes.

$$G_{k+1}(x) = 0 \ G_k(x) \qquad\qquad 0 \le x < 2^k$$
$$G_{k+1}(x) = 1 \ G_k(2^{k+1} - 1 - x) \qquad 2^k \le x < 2^{k+1}$$

E 5. Why does the Chinese ring puzzle solution have less than 2^k steps?

P 6. Using the two rules given for operation of the ring puzzle, write a recursive program to generate the sequence of steps to remove all rings from an initial configuration where all rings are on the bar.

E 7. Prove, by induction, that the recursive solution to the Towers of Hanoi puzzle takes $2^n - 1$ steps to move n disks.

P 8. Given the Gray code position and move corresponding to a Towers of Hanoi position, specify the Towers of Hanoi move.

P 9. Using the correspondence between the Towers of Hanoi solution and the Gray code solution, modify the nonrecursive Towers of Hanoi solution to obtain a nonrecursive Gray code generator.

E 10.   Define the term algorithm. According to your definition, can the two versions of the gcd algorithm given in the chapter be considered as examples of a single algorithm?

E 11.   Give an example of an algorithm for computing the gcd that you would not consider to be a version of Euclid's algorithm.

SUGGESTED PROGRAMMING EXERCISES

1.   Write a program corresponding to the flowchart of Figure 0-2. Modify the program so that it checks the assertion given in the text that "after k traversals of the loop, the value of x is $2k + 1$."

2.   Modify the Towers of Hanoi programs given in the text so that they generate a printout of the sequence of moves made. Run both the recursive and nonrecursive programs and compare computation costs.

sets, functions, and relations

SUMMARY

The first major abstraction that we will develop is the concept of a set. The concept of a set allows us to talk about a collection of objects, ignoring any order and ignoring repetition. Each set may have many representations, just as each number may have many numeric representations (e.g., $(15)_{10} = (17)_8 = XV = 6 + 3 * 3$). Since the purpose of many programs is to merge or otherwise manipulate collections of objects, the set notion is important in programming, and different operations for generating and forming sets will be discussed.

A secondary theme is the role of sets as the essential building blocks of modern mathematics and, therefore, as a foundation for computer science. The primary emphasis is on the set operations that are known to be applicable to computer science, but there is an occasional brief digression to a more theoretical result (e.g., even though sets ignore order, they can be used to represent the idea of order—through a clever artifice).

The second major concept is the concept of a relation, which allows us to formalize the notion of relationships that exist among the sets under discussion (e.g., given a set of objects and a set of colors, a relation may be used to describe the colors of the different objects). Ways of combining relations are considered, and properties of relations are developed. Particular combinations of these properties yield special kinds of relations, such as equivalence relations, which generalize the idea of equality, and partial orders, which generalize the notion of a linear order. Also of considerable importance is the relational description of data bases, which is very briefly introduced.

The third major concept is functionality. A function, formally defined as a relation with certain properties, is an abstraction of a rule assigning unique elements from one set to elements of another. In stating that there is a function that relates the sets, we do not necessarily specify how to compute it. In fact, in Chapter 4, it is shown that there are functions that no program can compute; they are known as noncomputable functions and are important because they help us to clarify the limitations of the algorithmic method.

Finally, the connection between relations and functions is discussed; the fact that each relation may be described as a

special kind of function is explained. This has important im-
plications for understanding programs that do not compute a
unique result, which are called nondeterministic programs.

A. SETS AND SEQUENCES

Definition
    A set *is a collection of distinct objects, without repetition,*
and without ordering.

The elements of a set are referred to as its *members.* We
sometimes write $A \in S$ to indicate that the element A is a
member of the set S, and $A \notin S$ for A is not a member of S.

Examples
    Days in the week = {Tuesday, Thursday, Sunday, Saturday,
Friday, Wednesday, Monday, Tuesday}.
    (Note that the listing of a set may contain repetitions, but
the set itself does not.)
    Set of positive integers less than $5 = \{1, 2, 3, 4\} = \{1, 4, 3, 2\} =$
$\{1, 1, 2, 3, 2, 4, 3, 3\}$.
    (Note that sets are equal when they have the same
members.) #
    An alternative method of defining particular sets is by a
description of some attribute or characteristic of the elements of
the set; in this case, the set is said to be *implicitly* specified.

Examples
    $S = \{x \mid x$ has property $p\}$.
S is the set of all elements having property p.
    $\{x \mid x$ is a three-letter word starting with c and ending with
$t\} = \{cat, cot, cut\}$.
    $\{x \mid x$ is an integer and $0 < x < 5\} = \{1, 2, 3, 4\}$.
    $S_1 = \{x \mid x = y^2$ for some positive integer $y\} = \{1, 4, 9, 16, \ldots\}$.
S_1 cannot be listed explicitly, since it is an infinite set.
    $S_2 = \{x \mid x$ is a valid Fortran identifier}.
    S_2 describes a finite set of strings.[1] #

---

[1]A *string* is a *sequence* of characters; "sequence" is defined in the text
on p. 26.

One type of implicit specification defines the members of a set by a computation rule for calculating members of the set; in this case the set is said to be *recursively* specified.

Example
Let $a_0 = 1$, $a_1 = 1$,   and $a_{i+1} = a_i + a_{i-1} i \geq 1$,   and $S = \{a_i | i \geq 0\}$.   #

Definition
A sequence (*or linear* array) *is a collection of elements listed in a linear order and having a first member.*

An element listed in a sequence can be referred to by a natural number that designates its position in the sequence. (The set of natural numbers consists of all positive integers and 0.)

Example
$$\mathbf{A} = (1, 2, 3, 5, 8, 13, 21).$$

The sequence **A** has seven members. One is the 0th element of the sequence **A**, two is the 1st element of **A**, . . . , 21 is the 6th element of **A**. The position of an element in a sequence is called its *index.*   #
Two sequences are equal just in case the members in corresponding positions in the two sequences are equal.

Example
$\mathbf{A}' = (1, 3, 2, 8, 5, 13, 21)$ and $\mathbf{A}'' = (1, 1, 3, 2, 8, 5, 13, 21)$.

Then $\mathbf{A}' \neq \mathbf{A}$, $\mathbf{A}' \neq \mathbf{A}''$, $\mathbf{A} \neq \mathbf{A}''$.
Note that if we had considered sets

$$A = \{1, 2, 3, 5, 8, 13, 21\}$$
$$A' = \{1, 3, 2, 5, 8, 13, 21\}$$
$$A'' = \{1, 1, 3, 2, 5, 8, 13, 21\}$$

it would be true that $A = A' = A''$.   #
If **S** is a sequence, set (**S**) is taken to be the set of all elements in the sequence **S**. Clearly, from our examples, set (**S**) = set (**S**') does not imply that **S** = **S**'.

Fact.   Given any sequence **S**, the corresponding set, set (**S**), can always be constructed.

But.   Given any set, S, it is not always possible to find a sequence **S** so that set (**S**) = S. When a sequence **S** can be found, the set S is said to be *enumerable*.

Fact.   Every finite set is enumerable.

Example

The set of real numbers, $0 \leq x \leq 1$, is not enumerable. Each number between 0 and 1 can be written as a decimal fraction $(0.999 \ldots = 1)$. Suppose that the sequence **S** of decimal fractions is presented, and it is claimed that every decimal fraction between 0 and 1 is a member of **S**. Let S_i be the ith element of **S**, and let S_{ij} be the jth digit of S_i. (Note that each S_i can be represented as an infinite sequence with zeros added to the right if necessary.) Arrange the sequences as shown in Figure 1-1.

Referring to Figure 1-1, a fraction t (in decimal form) is constructed by choosing the ith digit of t as follows: $t_i = S_{i,i} + 1$ if $S_{i,i}$ is not 8 or 9, and $t_i = 7$ if $S_{i,i} = 8$ or $S_{i,i} = 9$. t chosen in this way differs from the nth element of **S**, S_n, in the nth digit place. Thus t differs in at least one digit with each number in **S**, and t cannot have been in the sequence **S**. But if t is not included in **S**, then **S** is not an enumeration of all the numbers between 0 and 1. Since the only thing assumed about **S** is that it was an enumeration of the real numbers between 0 and 1, and this leads to a contradiction, we may conclude that the set S is not enumerable.   #

Since the preceding proof utilizes a construction that changes every element along the diagonal, it is called a proof by diagonalization.

$S_{0,0}$	$S_{0,1}$	$S_{0,2}$	$\cdots\cdots$	$S_{0,i-1}$	$S_{0,i}$	$\cdots\cdots$
$S_{1,0}$	$S_{1,1}$	$S_{1,2}$	$\cdots\cdots$	$S_{1,i-1}$	$S_{1,i}$	$\cdots\cdots$
$\cdot\cdot$	$\cdot\cdot$	$\cdot\cdot$		$\cdot\cdot$	$\cdot\cdot$	
$\cdot\cdot$	$\cdot\cdot$	$\cdot\cdot$		$\cdot\cdot$	$\cdot\cdot$	
$\cdot\cdot$	$\cdot\cdot$	$\cdot\cdot$		$\cdot\cdot$	$\cdot\cdot$	
$S_{i,0}$	$S_{i,1}$	$S_{i,2}$		$S_{i,i-1}$	$S_{i,1}$	$\cdots\cdots$
$\cdot\cdot$	$\cdot\cdot$	$\cdot\cdot$		$\cdot\cdot$	$\cdot\cdot$	
$\cdot\cdot$	$\cdot\cdot$	$\cdot\cdot$		$\cdot\cdot$	$\cdot\cdot$	

Figure 1-1.   Cantor diagonalization.

On proofs: The notion of a proof is one that is used both in an informal sense and, in Chapter 4, in a more formal sense. In the informal sense, a proof is nothing more than a convincing argument and is, of course, subject to our human fallibility. In the formal sense, a proof is a description of a convincing argument that is subject to algorithmic checking.

There are different strategies of proof that correspond to different modes of thinking. Furthermore, some proofs are difficult to follow because of the subtlety of the concepts or the length of the proof; in such cases, simpler proofs are sought to assure ourselves that the asserted conclusions do follow. Some of the common proof strategies seen so far include:

Proof by induction: This style of proof was presented in Chapter 0. It is generally used to prove a property of the integers. Its generalization, proof by recursion, will be discussed in Chapter 5.

Proof by contradiction: The diagonalization proof just given is an example of this style of proof. In the proof, an erroneous assumption is deliberately introduced and is shown to lead to a contradiction, proving that the hypothetical assumption was, in fact, wrong. In the diagonalization proof, the erroneous assumption is that the real numbers between 0 and 1 can be arranged in sequence.

As each new proof is encountered, interested students may try to analyze the pattern of the proof and see what other proof methods might be used to derive the same result. Important results are often confirmed by having several different lines of reasoning. For example, the Pythagorean theorem has been proved in hundreds of different ways.

The more general concept of a structure will be introduced in Chapter 5. The ordered set of even integers, $E = \langle \ldots, -4, -2, 0, 2, 4, \ldots \rangle$, is not a set because it is ordered. E is not a sequence because it has no first member. E is a structure.

B. SET CONTAINMENT

Definition

Set A is said to be contained in set B if every member of A is a member of B. We also say that A is a subset of B, denoted $A \subset B$.

The difference between membership and containment may perhaps be best understood at first considering the set symbols, { }, as a box. Then {a} is "a box holding a," while {{a}} is "a box holding a box holding a." The only subsets of {a} are {a} and { }; the subsets of {{a}} are {{a}} and { }.

As another illustration of the difference between membership and containment, consider the set of students taking this course. Each student taking this course is a member of the set. The set of students taking the course who are left-handed is a subset of the whole set.

If $A \subset B$, it is possible that $A = B$. If we wish to emphasize this fact, we write $A \subseteq B$. If equality is to be precluded, we write $A \subsetneqq B$, and A is said to be *properly* contained in B or, alternatively, A is a *proper* subset of B. As noted, two sets are equal just in case their members are the same. (*Note.* The phrase "just in case" is often used in proofs as a stylistic variant of the phrase "if and only if.")

Fact. $A = B$ if and only if $A \subseteq B$ and $B \subseteq A$.

Proof.   If $A = B$, by the definition of containment, $A \subseteq B$ and $B \subseteq A$, since every member of one is in the other.

Next suppose that $A \neq B$. Then there is either an element of A that is not in B or there is an element of B that is not in A. But $A \subseteq B$, so every element of A is in B, and $B \subseteq A$. So it cannot be that $A \neq B$.   #

Discussion of Proof.   This fact is of the form P if and only if Q. Thus there are two subfacts to prove: if P then Q; and P only if Q. The second subfact "P only if Q," may be paraphrased as "if Q then P." The first subfact is of the form "if P then both R and S" and is shown by demonstrating the two sub-subfacts "if P then R" and "if P then S" which, in this case, are "if $A = B$ then $A \subset B$" and "if $A = B$ then $B \subset A$." The second subfact, "if Q then P," is shown by an argument known as the *contrapositive*, in which it is shown that not P implies not Q, from which "if Q then P'' follows. Again, in this case, this is of the form "if R and S then P," and the contrapositive proof shows that not P implies either not R or not S.

Example
   $\{1, 2, 3\} \subseteq \{1, 2, 3, 4\}$, but $\{1, 2, 3\} \neq \{1, 2, 3, 4\}$. $\{1, 2, 3, 4\} = \{1, 3, 2, 4, 3\}$, since $\{1, 2, 3, 4\} \subseteq \{1, 3, 2, 4, 3\}$ and $\{1, 3, 2, 4, 3\} \subseteq \{1, 2, 3, 4\}$. #
Fact. Every set is a *subset* of itself, but no set is a *member* of itself.

   The implicit definition of sets does not allow a set to be a member of itself, since without this restriction we are led to Russell's paradox.

Let R be the set of all sets that are not members of themselves; that is, $R = \{S \mid S \notin S\}$.

Now let us see if R is a member of R. First suppose that $R \notin R$; by the definition of R as the set of all sets that are not members of themselves, R is indeed in R. Thus, $R \notin R$ implies $R \in R$. Next suppose that $R \in R$; again, by the definition of R, it must have the property that it is not a member of itself. Thus $R \in R$ implies $R \notin R$. From this information we are led to the paradoxical conclusion that $R \in R$ if and only if $R \notin R$.

   The elements of a set may themselves be sets, sequences, or structures.

Examples
   Let $S_1 = \{a\}$ and $S_2 = \{\{a\}\}$. Then $S_1 \neq S_2$. S_1 and S_2 have no members in common; each has only one member. Let $S_3 = \{a, \{a\}\}$. Then S_3 has two members. $S_3 \neq S_2$ and $S_3 \neq S_1$. But $S_1 \subseteq S_3$ and $S_2 \subseteq S_3$. (Also, S_1 is a member of S_3.) #
   An important result is that there are many questions about implicitly specified sets that are not answerable. For example, if A and B are implicitly specified, it may be impossible to determine if $A \subseteq B$. When it can be shown that no well-defined procedure can be given for a problem, the problem is said to be algorithmically *unsolvable*. If a "yes-no" question is unsolvable, then this particular question is algorithmically *undecidable*.
   The notion of a question being undecidable is that there is

no algorithm to solve all cases of the general question, but particular instances or special cases may have solutions. This concept is so important in computer science (and is possibly so novel to students) that it is introduced here. More discussion on this point is contained in Chapter 4.

Perhaps the following true story will help drive home the point. When I first began teaching computer science, I gave the following examination question. Discuss the difficulty of deciding whether two specifications of a language are equivalent. Imagine my surprise several months later when I learned that a diligent student had been trying valiantly since the exam to program a system to test the equivalence of specifications. Of course, such an attempt is foredoomed, since no algorithm can decide the equivalence of specifications. It's undecidable! The moral is: Beware of the limits of the algorithmic method!

Examples

Many times containment and membership are undecidable. Let F be the flowchart of Figure 1-2, where $P(X)$ is an arbitrary

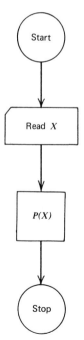

Figure 1-2. Flowchart of a program of one variable.

program in some procedure-oriented language. Let S be the set of values of X for which F stops.

For arbitrary fixed P and X, it is undecidable if X is in S.

For arbitrary P_1 and P_2, if S_i is the set of argument values for which F stops when executing P_i, then it is undecidable whether $S_1 \subseteq S_2$. #

Of course, in many cases it is possible to decide questions of containment or membership.

Example

$S_1 = \{n^2 \mid n$ is an odd integer$\}$ and $S_2 = \{n \mid n$ is an odd integer$\}$. Then it is easy to show that $S_1 \subseteq S_2$.

(a) 3 is in S_2 but not in S_1, so $S_1 \neq S_2$.

(b) Let k be an element of S_1. Then $k = n^2$ for some odd integer $n = 2j + 1$. So $k = (2j + 1)^2 = 4j^2 + 4j + 1 = 2(2j^2 + 2j) + 1$. Thus k is odd and is therefore a member of S_2. Since k was arbitrary, every element of S_1 is also an element of S_2 and $S_1 \subseteq S_2$. #

But, in general, the following questions are also undecidable about implicitly specified sets.

1. Does S have any members?

2. Does S have a finite number of members?

3. Is some particular x included in S? #

C. THE EMPTY SET

Definition

An empty set, *denoted* $\emptyset$ or $\{\ \}$, *is a set with no members.*

If $\emptyset_1$ and $\emptyset_2$ are empty sets, then $\emptyset_1 \subseteq \emptyset_2$ and $\emptyset_2 \subseteq \emptyset_1$. (Why?) Therefore, $\emptyset_1 = \emptyset_2$. Thus there is exactly one empty set, $\emptyset$.

Examples

$$\emptyset = \{x \mid 2x = 2x + 1\}$$
$$\emptyset = \{x \mid x < 4 \text{ and } x > 3 \text{ and } x \text{ is an integer}\} \quad \#$$

If S is any set, then $\emptyset \subseteq S$. The ubiquitous empty set is a subset of every set. The difference between membership and containment is illustrated by $\emptyset$. For any set S, $\emptyset \subseteq S$ but, in general, $\emptyset \notin S$.

Examples
    If $S = \{\emptyset\}$, $\emptyset \subseteq S$ and $\emptyset \in S$.
    If $S = \{\{\emptyset\}\}$, $\emptyset \subseteq S$ but $\emptyset \notin S$.   #

As noted, given an implicitly specified set, S, it is generally algorithmically undecidable if $S = \emptyset$.

D. THE POWER SET

Definition
    The set of all subsets of S is called the *power set of S*, denoted $P(S)$ or 2^S; $P(S) = \{X \mid X \subseteq S\}$.

Examples
    If $S = \{1, 2, 3\}$, $P(S) = \{\emptyset, \{1\}, \{2\}, \{3\}, \{1, 2\}, \{1, 3\}, \{2, 3\}, \{1, 2, 3\}\}$. Note that S has 3 members and $P(S)$ has 2^3 members. In general, if S has k members, $P(S)$ has 2^k members.

$$\emptyset = \{\ \}$$
$$P(\emptyset) = \{\emptyset\}$$
$$P(P(\emptyset)) = \{\emptyset, \{\emptyset\}\}$$
$$P(P(P(\emptyset))) = \{\emptyset, \{\emptyset\}, \{\{\emptyset\}\}, \{\emptyset, \{\emptyset\}\}\}   \#$$

In the programming language Pascal the notion of power set is used to define a data type in the language. The set of colors may be defined by declaring TYPE primary color = (BLUE, YELLOW, RED). Then TYPE color = *set of* primary color. In this declaration, the objects of type color are subsets of the set of primary colors and, therefore, members of the power set of the set of primary colors.

FACT. A set is never equal to its power set, since a set is a member of its power set but cannot be a member of itself. We prove this later in Section M.

Starting with $\emptyset$ as a representation of the number 0, one can construct a representation of the set of positive integers in which a number is represented as $P^k(\emptyset)$, where $P^0(\emptyset) = \emptyset$ and

$P^{i+1}(\emptyset) = P(P^i(\emptyset))$. Thus

$$P^0(\emptyset) = \emptyset \leftrightarrow 0$$
$$P^1(\emptyset) = P(\emptyset) \leftrightarrow 1$$
$$P^2(\emptyset) = P(P^1(\emptyset)) = P(P(\emptyset)) \leftrightarrow 2$$
$$P^3(\emptyset) = P(P^2(\emptyset)) = P(P(P(\emptyset))) \leftrightarrow 3$$

etc.

Definition
    The cardinality *of a set, S, is the number of elements that are members of the set, S, and is denoted* $|S|$.

Example
    $S = \{1, 2, 3\}$, $|S| = 3$, and $|P(S)| = 2^3 = 8$.   #
    In Section M it is shown that a set, S, and its power set, $P(S)$, can never be of the same size. The power set is always larger, since there is a subset of the power set that is of the same size as the original set: $Q = \{\{x\}|\, x \in S\}$, or Q is the subset of $P(S)$ consisting of those sets that are singletons. The set of natural numbers, N, is infinite, and any other set of the same size is said to be *countable*. The set of subsets of N, $P(N)$, is larger than N, $|P(N)| > |N|$; taking successive power sets $P(P(N))$, $P(P(P(N)))$, we can get larger and larger *uncountable* sets.

E. OPERATIONS ON SETS
    The power set operation P generating the power set is a unary operation on S, forming a new set from a single given set. Binary operations form new sets from pairs of sets.

Definition
    The union *of sets A and B is the smallest set containing all members of A and all members of B, denoted* $A \cup B$. *Thus* $A \cup B = \{x|\, x$ *is in A or x is in B*$\}$, *where the "or" is, as always, an inclusive or; that is, a or b means a or b or both ("or both" being redundant).*

Example
    $\{1, 2, 3, 5\} \cup \{1, 2, 3, 4\} = \{1, 2, 3, 4, 5\}$.   #
    In visualizing set operations, the Venn diagrams are helpful

for up to three sets. A Venn diagram for one, two, and three sets, respectively, is shown in Figure 1-3.

In Figure 1-3a, anything in A is in the circle and anything not in A is outside of the circle. In Figure 1-3b, anything in circle A is in set A and anything not in circle A is not a member of A. The total area is divided into four mutually exclusive areas, some of which may be empty.

Those elements in A but not in B.

Those elements both in A and in B.

Those elements in B but not in A.

Those elements in neither A nor B.

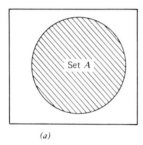

(a)

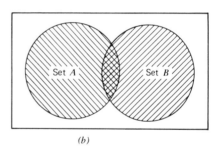

(b)

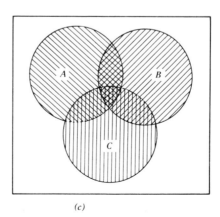

(c)

Figure 1-3.   Venn diagrams. (a) Single set. (b) Two sets. (c) Three sets.

Facts. $A \cup A = A$ idempotency
$A \cup B = B \cup A$ commutativity
$A \cup (B \cup C) = (A \cup B) \cup C$ associativity

All these facts follow from the observation that the Venn diagram areas spanned by the left-hand and right-hand sides of the equations are the same.

Definition

The intersection of sets A and B, denoted $A \cap B$, is the smallest set containing all the elements that are members of both A and B. Thus, $A \cap B = \{x \mid x$ is in A and x is in $B\}$.

Facts. $A \cap A = A$ idempotency
$A \cap B = B \cap A$ commutativity
$(A \cap B) \cap C = A \cap (B \cap C)$ associativity

The operations of union and intersection may be combined in a single expression. The following facts can be established from the Venn diagram.

Facts. $A \cap (B \cup C) = (A \cap B) \cup (A \cap C)$ left distributivity of intersection over union

$(A \cup B) \cap C = (A \cap C) \cup (B \cap C)$ right distributivity of intersection over union

$A \cup (B \cap C) = (A \cup B) \cap (A \cup C)$ left distributivity of union over intersection

$(A \cap B) \cup C = (A \cup C) \cap (B \cup C)$ right distributivity of union over intersection

Definition

The complement of B with respect to A, denoted $A - B$, is the set containing the elements of A that are not in B. Thus $A - B = \{x \mid x$ is in A and x is not in $B\}$. This operation is also called the relative complement.

Facts. $A - A = \emptyset$.

The relative complement is not a commutative or associative operation.

Examples

$A = \{1, 2, 3\}$, $B = \{2, 3, 4\}$, and $C = \{3, 4, 5\}$.

Then $A - B = \{1\}$ and $B - A = \{4\}$, so that $A - B \neq B - A$, showing that relative complement is not commutative.

Next, $(A - B) - C = \{1\}$, while $A - (B - C) = \{1, 3\}$. This shows that relative complement is not associative. #

Fact. $A - (A - B) \subseteqq B$.

Definition

The symmetric difference *of A and B is denoted* $A + B$ *or* $A \oplus B$ *and is defined by* $A + B = (A - B) \cup (B - A)$. *Thus* $A + B = \{x \mid x$ *is in A but not in B or x is in B but not in A*\}.

Facts. $A + A = \emptyset$, $A + B = B + A$, and $A + (B + C) = (A + B) + C$.

Definition

If A is understood as a given universe, then $A - B$ *is referred to simply as the* complement *of B, denoted* $-B$ *or* $\bar{B}$.

Fact. $A - B = A \cap \bar{B}$.

Fact. $\bar{\bar{A}} = A$. (A double negative is positive!)

DEMORGAN'S LAWS. $\overline{A \cup B} = \bar{A} \cap \bar{B}$ and $\overline{A \cap B} = \bar{A} \cup \bar{B}$.

DeMorgan's laws can be established from the Venn diagram. If the area outside A represents $\bar{A}$ and the area outside B represents $\bar{B}$, the proof is immediate.

We may also proceed algebraically to establish relationships among set expressions. Let I be our universe; then, given sets A, B we have $A \cup \bar{A} = B \cup \bar{B} = I$, and $I = I \cap I = (A \cup \bar{A}) \cap (B \cup \bar{B}) = (A \cap B) \cup (A \cap \bar{B}) \cup (\bar{A} \cap B) \cup (\bar{A} \cap \bar{B})$. Thus the universe, I, is the union of the four disjoint sets $A \cap B$, $A \cap \bar{B}$, $\bar{A} \cap B$, and $\bar{A} \cap \bar{B}$.

Two sets A and B are said to be disjoint if $A \cap B = \emptyset$. If $S_1 = S_2 \cup S_3$ and S_2, S_3 are disjoint, $S_2 = S_1 - S_3$. Conversely, if $S_2 = S_1 - S_3$ and $S_3 \subset S_1$, S_2 and S_3 are disjoint and $S_1 = S_2 \cup S_3$.

Applying DeMorgan's laws, $A \cup B$ can be expressed as a union of disjoint sets: $A \cup B = (\overline{\bar{A} \cap \bar{B}}) = I - (\bar{A} \cap \bar{B}) = (A \cap B) \cup (A \cap \bar{B}) \cup (\bar{A} \cap B)$.

Example

$$A - (A - B) = A - (A \cap \bar{B}) \qquad \text{definition of } X - Y$$
$$= A \cap \overline{(A \cap \bar{B})} \qquad \text{definition of } X - Y$$
$$\text{where } Y \text{ is } (A \cap \bar{B})$$
$$= A \cap (\bar{A} \cup B) \qquad \text{DeMorgan's law}$$
$$= (A \cap \bar{A}) \cup (A \cap B) \qquad \text{distributive law}$$
$$= \emptyset \cup (A \cap B) \qquad X \cap \bar{X} = \emptyset$$
$$= A \cap B \qquad \emptyset \cup X = X \quad \#$$

Example

We can now show, algebraically, that $A - (B - C) \neq (A - B) - C$: $A - (B - C) = A \cap \overline{(B \cap \bar{C})} = A \cap ((B \cap C) \cup (\bar{B} \cap C) \cup (\bar{B} \cap \bar{C})) = (A \cap B \cap C) \cup (A \cap \bar{B} \cap C) \cup (A \cap \bar{B} \cap \bar{C})$, and $(A - B) - C = A \cap \bar{B} \cap \bar{C}$. Now $A \cap B \cap C$, $A \cap \bar{B} \cap C$, and $A \cap \bar{B} \cap \bar{C}$ are pairwise disjoint. (Why?) Therefore $(A - B) - C = A - (B - C)$ just in case $(A \cap B \cap C) \cup (A \cap \bar{B} \cap C) = \emptyset$. We can simplify this condition as follows.

$$(A \cap B \cap C) \cup (A \cap \bar{B} \cap C)$$
$$= (A \cap C \cap B) \cup (A \cap C \cap \bar{B}) \qquad \text{commutativity}$$
$$= (A \cap C) \cap (B \cup \bar{B}) \qquad \text{distributivity}$$
$$= (A \cap C) \cap I \qquad \text{since} \quad B \cup \bar{B} = I$$
$$= (A \cap C) \qquad \text{since} \quad X \cap I = X \quad \text{for all } X$$

Thus $A - (B - C) = (A - B) - C$ if and only if $A \cap C = \emptyset$. #

Example

$$A = \{1, 2, 3\}, \qquad B = \{3, 4, 5\}, \qquad \text{and} \qquad C = \{5, 6, 7\}.$$
In this case $A \cap C = \emptyset$.

$$(A - B) = \{1, 2, 3\} - \{3, 4, 5\} = \{1, 2\}$$
$$(A - B) - C = \{1, 2\} - \{5, 6, 7\} = \{1, 2\}$$
and
$$(B - C) = \{3, 4, 5\} - \{5, 6, 7\} = \{3, 4\}$$
$$A - (B - C) = \{1, 2, 3\} - \{3, 4\} = \{1, 2\} \qquad \#$$

Describing Abstract Algorithms

The set concept is useful because it allows us to concentrate on extensionality and ignore order. Descriptions of algorithms that are free of irrelevant detail are more understandable than programs in which a vast amount of detail is presented all at once because of the inability of the human mind's short-term memory to retain and operate on large amounts of data. Thus, descriptions that abstract and simplify are desirable for programs. The set concept is useful precisely because it allows us to concentrate on extensionality and ignore order.

To see this, the following abstract program is a presentation of the classical sieve of Eratosthenes for computing the set of prime numbers less than or equal to n.

$I := \{1, 2, \ldots, n\}$;
$I_0 := I - \{1\}$;
Primes $:= \emptyset$;
while $I_0 \neq \emptyset$ do
 begin new prime $:= \min(I_0)$;
 $I_0 := I_0 - $ new prime $*I$;
 comment new prime $*I$ is the set derived by multiplying each element of I by new prime end comment
 Primes $:=$ Primes $\cup \{$new prime$\}$
 end

Remarks to Skeptics

Having taken as an operating definition of a proof any convincing argument (of which the author, or editor, is the judge), it is perhaps still best to reply to skeptics who question the use of Venn diagrams as proofs. A Venn diagram argument is a demonstration that two set expressions, E_1 and E_2, define the same regions on the Venn diagram.

Example
 $x \cap \bar{y}$ and $(\overline{\bar{x} \cup y})$. #

The argument is *sound*, since two expressions defining the same region on the Venn diagram denote the same set. The argument is *complete*, since two expressions denoting the same set always define the same region on the Venn diagram. Furthermore, the Venn diagram argument is readily formalized and mechanized for any given finite number of sets. The justification of the Venn diagram argument and its formalization are given in Chapter 4;

by the time Chapter 4 is studied, the algebraic "machinery" and logical concepts will have been developed.

F. ORDER

An essential aspect of the definition of a set is the absence of any ordering of its elements. The concept of a set is purely extensional, and equality of sets is determined solely by their extents (i.e., members). However, it is possible to *represent* a sequence using sets.

Note.   The representation of ordered pairs as sets is one of the results that illustrate the surprising subtlety of the mathematical concepts. Although sets have been defined to abstract from the concept of order, they still retain the implicit capability to describe order. So this somewhat surprising aspect is presented, even though we know of no application of this fact.

Consider (a, b), a sequence of two elements, not necessarily distinct, that we want a set to represent in such a way that we can uniquely reconstruct the sequence given the set. $\{a, b\}$ will not do, since $\{a, b\} = \{b, a\}$ and does not distinguish (a, b) from (b, a), although these are distinct sequences.

The representation of (a, b) as a set is taken to be $\{\{a\}, \{a, b\}\}$; then (b, a) is represented as $\{\{b\}, \{b, a\}\}$.

Definition

The ordered pair (a, b) *is the sequence of two elements* (a, b).

Fact.   The set $\{\{a\}, \{a, b\}\}$ uniquely represents the ordered pair (a, b). Given an ordered pair, the set can be reconstructed, and given a set S, which represents an ordered pair, exactly one ordered pair will be represented by S.

Example

$$(1, 2) \leftrightarrow \{\{1\}, \{1, 2\}\}$$
$$(\emptyset, a) \leftrightarrow \{\{\emptyset\}, \{\emptyset, a\}\}$$
$$(a, b) \leftrightarrow \{\{a\}, \{a, b\}\}$$
$$(a, a) \leftrightarrow \{\{a\}, \{a, a\}\} = \{\{a\}, \{a\}\} = \{\{a\}\} \quad \#$$

The extension to sequences of length larger than 2 is usually done as follows:

for sequences of length 3: $(a, b, c) \leftrightarrow ((a, b), c)$

so that a sequence of length 3 is taken to be the same as a sequence of length 2 whose first member is a sequence of length 2.[2]

For sequences of length greater than 3, we have, in general,

$$S_n = (t_0, t_1, \ldots, t_{n-1}) = (S_{n-1}, t_{n-1}) \qquad n \geqq 3$$

We do not distinguish between (a, b, c, d) and $(((a, b), c), d)$, and we treat these as equal in computer representation.

G. RELATIONS

Relationships among data often are to be described or operated on. To formalize the intuitive notion of a relation, we first need the following information.

Definition

Let A, B be nonempty sets, and let $A \times B$ (read "A cross B") be defined as $A \times B = \{(a, b) | a$ is in A, and b is in $B\}$; that is, $A \times B$, the Cartesian product *of A and B, is the set of all ordered pairs (a, b) such that the first element of the ordered pair, a, is from A and the second element of the ordered pair, b, is from B.*

Example
  $\{1, 2\} \times \{a, b, c\} = \{(1, a), (1, b), (1, c), (2, a), (2, b), (2, c)\}$.   #

Now a *relation*, R, with *domain A* and *range B*, is any subset of $A \times B$. If (a, b) is in R, we often write aRb.

Notation. An alternative notation for ordered pairs that is often convenient, particularly in representing relations, is $\frac{a}{b}$ for the ordered pair (a, b).

Relations are sets, and many set operations can be applied to relations to obtain new relations. Thus, if R_1 and R_2 are two relations with the same domain, D, and the same range, R, then

---

[2]Note that in Lisp the reverse definition is taken: $(a, b, c) \leftrightarrow (a, (b, c))$. By symmetry, this works equally well.

$R_1 \cup R_2$ and $R_1 \cap R_2$ are relations with domain D and range R. Likewise, the relative complement $(D \times R) - R_1$ is a relation with domain D and range R.

Examples

Let domain = range = $\{1, 2, 3, 4\}$.

$$R(<) = \{(1, 2), (1, 3), (1, 4), (2, 3), (2, 4), (3, 4)\}$$
$$R(=) = \{(1, 1), (2, 2), (3, 3), (4, 4)\}$$
$$R(\leqq) = R(<) \cup R(=) \quad \#$$

Fact. For any specified nonempty domain and range, $\emptyset$ is a relation, called the *empty relation*. (Note that since $\emptyset$ is empty it has no members that are not ordered pairs from the specified domain and range.)

Recall that $A \times B = \{(a, b)|$ a is in A, b is in $B\}$ is the Cartesian product of sets A and B. This generalizes directly as follows.

Definition

$A_0 \times A_1 \times \cdots \times A_{n-1}$ *is the set of all sequences of length n,* $(a_0, a_1, \ldots, a_{n-1})$, *where a_0 is in A_0, a_1 is in $A_1, \ldots$, and a_{n-1} is in A_{n-1}.*

Example

Let A be a finite alphabet. Then $A \times A \times A \times A \times A \times A$ is the set of all six letter sequences over alphabet A.

Relational Data Bases

Definition

A subset of the n-fold Cartesian product, $R_n \subseteq D_1 \times D_2 \times \cdots \times D_n$, is called an n-ary relation. An element of R_n is a list of n-elements, $(d_1, d_2, \ldots, d_n)$.

In the relational model of data bases such *n*-ary relations are used as the standard method for describing the relationships in the data.

Example

R_1:	Course Number	Section Number	Instructor	Course Title
	CS 105	10	Levy	General Computer Science
	CS 105	20	Khalil	General Computer Science
	CS 105	30	Cassel	General Computer Science
	CS 170	10	Carberry	Computer Science I
	CS 170	20	Leathrum	Computer Science I
	CS 240	10	Levy	Discrete Structures
	CS 240	20	Slutzki	Discrete Structures

In the relation R_1, each row is a distinct element of the relation. Furthermore, the ordering of the rows in the table is not significant.

The relations in a data base are a description of the data, but different ways of describing the same data are often available. In transforming the data representation, one useful operation for forming new relations from a given relation is *projection*. #

Definition

Let R_n be an n-ary relation, $R_n \subseteq D_1 \times D_2 \times \cdots \times D_n$; the projection of R_n onto $D_{i_1}, D_{i_2}, \ldots, D_{i_m}$ is an m-ary relation, obtained from R_n by choosing the i_1th, i_2th, ..., i_mth components of each element of R_n. We denote this projection $\pi R_{nD_{i_1}, D_{i_2} \ldots D_{i_m}}$.

Example

In the relation R_1, the projection of R_1 onto course number, section number, and instructor is:

R_2:	Course Number	Section Number	Instructor
	CS 105	10	Levy
	CS 105	20	Khalil
	CS 105	30	Cassel
	CS 170	10	Carberry
	CS 170	20	Leathrum
	CS 240	10	Levy
	CS 240	20	Slutzki

The projection of R_1 onto course number and course title, is:

R_3:	Course Number	Course Title
	CS 105	General Computer Science
	CS 170	Computer Science I
	CS 240	Discrete Structures

#

R_1 can be reconstructed from R_2 and R_3, so that R_2 and R_3 are just relational representations of the same data that are contained in R_1.

The theory of relational data bases is concerned with the different ways of representing data relations, and the operations needed to query and update the data base.

H. PROPERTIES OF RELATIONS

Relations where the domain and range are identical often possess properties that describe some aspect of the relations and its applicability.

Definition
$R \subseteq A \times A$ is reflexive *if every element is related to itself.*

Examples
Equality is a reflexive relation, since every element equals itself (and so "is related by the equality relation").

The empty relation (over any nonempty set) is not reflexive, since no element is related to itself.

$R = \{(1, 1), (1, 2), (2, 2)\} \subseteq \{1, 2\} \times \{1, 2\}$ is reflexive because both $(1, 1)$ and $(2, 2)$ are in R.

Definition
$R \subseteq A \times A$ is irreflexive *if no element is related to itself; that is, (a, a) is not in R for any a in A.*

Definition
$R \subseteq A \times A$ is symmetric *if, whenever (a, b) is in R, (b, a) is also in R.*

Definition
$R \subseteq A \times A$ is asymmetric *if, whenever (a, b) is in R, (b, a) is not in R.*

Definition
$R \subseteq A \times A$ is antisymmetric *if, whenever (a, b) and (b, a) are in R, then a must equal b. [The only symmetric ordered pairs in R are of the form (a, a).]*

Definition
$R \subseteq A \times A$ is transitive *if, whenever (a, b) is in R and (b, c) is also in R, (a, c) is in R also.*

Examples
Let domain = range = $\{1, 2, 3\}$, and

$$R_1 = \{(1, 1), (1, 2), (2, 2), (2, 3)\}$$
$$R_2 = \{(1, 2), (2, 3), (1, 3)\}$$
$$R_3 = \{(1, 1), (2, 2), (2, 3), (3, 2), (3, 3)\}$$

R_2 is asymmetric, but R_1 and R_3 are not. [No element of the form (a, a) can appear in an asymmetric relation.) R_1 and R_2 are antisymmetric; R_3 is not. R_2 and R_3 are transitive; R_1 is not, since $(1, 2)$ and $(2, 3)$ are in R_1 but $(1, 3)$ is not.   #

Definition
If R is reflexive, symmetric, and transitive, R is said to be an equivalence relation.

Equivalence relations play a fundamental role in the anlysis of discrete structures; when a set of elements is related in an equivalence relation, each element of the set is related to every other element of the set and to itself. This is a direct consequence of the properties of *reflexivity, symmetry,* and *transitivity.*

Example
R_3 in the preceding example is an equivalence relation.   #
More concise notation for representing the properties of relations is developed in Section L.

I. PARTITIONS

Definition
    A partition *of the set A is a set B whose members are pairwise, disjoint, nonempty sets. The members of B are called blocks of the partition. Finally, every member of A is a member of some block of the partition.*

Example
    The partitions of $\{1, 2, 3\}$ are:

$$\{\{1\}, \{2\}, \{3\}\}$$
$$\{\{1, 2, 3\}\}$$
$$\{\{1\}, \{2, 3\}\}$$
$$\{\{2\}, \{1, 3\}\}$$
$$\{\{3\}, \{1, 2\}\} \quad \#$$

Notation. It is customary to use the following notation for partitions: blocks are written separated by semicolons.

Examples
    $\{\{a_1, a_2, a_3, \ldots\}, \{b_1, b_2, b_3, \ldots\}, \{c_1, c_2, c_3, \ldots\}\}$ is written $\{a_1, a_2, a_3, \ldots ; b_1, b_2, b_3, \ldots ; c_1, c_2, c_3, \ldots\}$,

$$\{\{1\}, \{2, 3\}\} \text{ is written } \{1; 2, 3\}$$
$$\{\{1\}, \{2\}, \{3\}\} \text{ is written } \{1; 2; 3\} \quad \#$$

Facts. Partitions and equivalence relations are related concepts. With each partition, π, may be associated an equivalence relation, R_π, in which $aR_\pi b$ if and only if a and b are in the same block of π. The fact that R_π is an equivalence relation is easy to verify from the definition, since R_π must be reflexive, symmetric, and transitive.
    Conversely, if R is reflexive, symmetric, and transitive, a partition, π_R, may be defined. In the same block of π_R as any element, a, are all the elements related to a by the equivalence relation R.

Examples
    Let $R = \{(1, 1), (1, 3), (2, 2), (2, 4), (3, 1), (3, 3), (4, 2), (4, 4)\}$. It may be verified that R is an equivalence relation. The correspond-

ing partition is $\pi_R = \{1, 3; 2, 4\}$. Let $\pi = \{1, 2; 3\}$. Then $R_\pi = \{(1, 1),$ $(1, 2), (2, 1), (2, 2), (3, 3)\}$.   #

An equivalence relation is a relation that is reflexive, symmetric, and transitive. We generalize this as follows.

Definition
A compatibility relation *is a relation on a set that is reflexive and symmetric.*

Example
$R \subseteq \{1, 2, 3, 4\} \times \{1, 2, 3, 4\}$

$R = \{(1, 1), (1, 2), (1, 3), (2, 1), (2, 2), (2, 3), (2, 4), (3, 1), (3, 2),$
$\quad (3, 3), (3, 4), (4, 2), (4, 3), (4, 4)\}.$

It is easily checked that R is a compatibility relation (do it), but that R is not transitive, since $(1, 2)$ and $(2, 4)$ are in R but $(1, 4)$ is not in R and, therefore, R is not an equivalence relation.   #

It follows directly from the definition of compatibility relation that every equivalence relation is a compatibility relation and, as the last example shows, not every compatibility relation is an equivalence relation.

Given a compatibility relation, R, we may define the complete cover for R as follows.

Definition
Let R *be a compatibility relation; a* complete cover *for R is a set $C = \{B_i\}$, where (a, b) is in R if and only if a and b are in some B_i and each B_i is maximal (i.e., any element related to all the members of any B_i is already in that B_i).*

Example
For the compatibility relation of the previous example, the complete cover is

$$\{\{1, 2, 3\}, \{2, 3, 4\}\}$$

No element can be added to either of the sets of the cover, since 1 and 4 are not related in R.   #

It is easy to see that if R is an equivalence relation, the complete cover for R is the partition associated with R.

Fact. Each compatibility relation, R, uniquely determines a complete cover.

The analogy between equivalence relations and compatibility relations can be seen as follows.

compatibility relation over A	complete cover of A	relation from A to "blocks" of the complete cover
equivalence relation over A	partition of A	rule assigning an element of A to a block of the partition

J. THE PRINCIPLE OF INCLUSION AND EXCLUSION

Another operation on sets is calculating the number of elements in a specified combination of finite sets.

Example

There are 50 students in the class. Ten have blue eyes, 15 have brown hair, and 23 are majoring in computer science. Eight have blue eyes and brown hair, 7 have blue eyes and are majoring in computer science, and 10 have brown hair and are majoring in computer science. Finally, 4 students have blue eyes, brown hair, and are majoring in computer science. How many students have blue eyes, or brown hair, or are majoring in computer science?   #

Notation.   Let R = set of brown haired students, L = set of blue eyed students, and C = set of computer science majors. If x is a given set, $N(x)$[3] denotes the cardinality of x. The data given can then be summarized in this notation as:

$$N(R) = 15 \qquad N(L) = 10 \qquad N(C) = 23 \qquad N(R \cap L) = 6$$
$$N(L \cap C) = 7 \qquad N(R \cap C) = 10 \qquad N(R \cap L \cap C) = 4$$

The problem is to determine $N(R \cup L \cup C)$.

Example (cont.)

This problem can be attacked by considering a Venn diagram.

---

[3]$N(x)$ is an alternative notation to $|x|$, which is more convenient in this case.

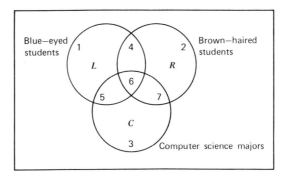

Let N_i denote the number of students in the elemental region of the diagram labeled i. Then the given data may be reexpressed as:

$$N(L) = N_1 + N_4 + N_5 + N_6 = 10 \qquad (1\text{-}1)$$

$$N(B) = N_2 + N_4 + N_6 + N_7 = 15 \qquad (1\text{-}2)$$

$$N(C) = N_3 + N_5 + N_6 + N_7 = 23 \qquad (1\text{-}3)$$

$$N(R \cap L) = N_4 + N_6 = 6 \qquad (1\text{-}4)$$

$$N(L \cap C) = N_5 + N_6 = 7 \qquad (1\text{-}5)$$

$$N(R \cap C) = N_6 + N_7 = 10 \qquad (1\text{-}6)$$

$$N(R \cap L \cap C) = N_6 = 4 \qquad (1\text{-}7)$$

From these equations the value of N_i can be determined for each i.

From Equations 1-4 to 1-7, $N_4 = 2$, $N_5 = 3$, and $N_7 = 6$. These data, together with Equations 1-1 to 1-3, determine

$$N_1 = 1, \qquad N_2 = 3, \qquad N_3 = 10$$

The distribution of students in the sets specified by each elemental region is as shown in the following Venn diagram.

It is easy to check that these data are consistent with the given data.   #

Although this method will work in cases where the reasoning can be applied directly to the Venn diagram, there is an algebraic approach that is often more convenient; it is known as the *principle of inclusion and exclusion*. The principle of in-

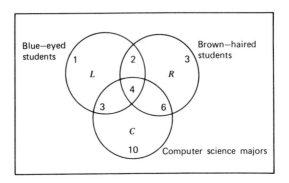

clusion and exclusion is based on the computation of the cardinality of the union sets. For two sets, A and B, $N(A \cup B) = N(A) + N(B) - N(A \cap B)$, which can be seen directly since, in counting $N(A)$ and $N(B)$, the elements in $N(A \cap B)$ have been counted twice—once in $N(A)$ and once in $N(B)$.

Example
Among the positive integers less than or equal to 100, let A be the set of even integers and let B be the set of integers divisible by 5. How many integers are even or divisible by 5?

The set whose cardinality is to be computed is $A \cup B$, while $N(A)$, $N(B)$, and $N(A \cap B)$ are readily seen to be

$$N(A) = 50, \qquad N(B) = 20, \qquad \text{and} \qquad N(A \cap B) = 10$$

Then

$$N(A \cup B) = N(A) + N(B) - N(A \cap B) = 50 + 20 - 10 = 60 \quad \#$$

The formula for the union of three sets is determined using the associativity of $\cup$.

$$N(A \cup B \cup C) = N(A \cup B) + N(C) - N((A \cup B) \cap C)$$
$$= N(A) + N(B) + N(C) - N(A \cap B) - N((A \cup B) \cap C)$$

The last term on the right can now be evaluated using the distributive law $(A \cup B) \cap C = (A \cap C) \cup (B \cap C)$ and the rule for evaluating unions.

$$N((A \cup B) \cap C) = N((A \cap C) \cup (B \cap C))$$
$$= N(A \cap C) + N(B \cap C) - N(A \cap B \cap C)$$

Thus

$$N(A \cup B \cup C) = N(A) + N(B) + N(C) - N(A \cap B) - N(A \cap C)$$
$$- N(B \cap C) + N(A \cap B \cap C)$$

The general principle of inclusion and exclusion can now be stated. (The proof is by induction and is omitted.)

$$N\left(\bigcup_{i=1}^{n} A_i\right) = \sum_{i=1}^{n} N(A_i) - \sum_{i \neq j} N(A_i \cap A_j) + \sum_{\substack{i,j,k \\ \text{all} \\ \text{different}}} N(A_i \cap A_j \cap A_k)$$

$$- \cdots + (-1)^{n-1} N(A_1 \cap A_2 \cap \cdots \cap A_n)$$

Example

Let x be the set of four-digit decimal integers. $x = \{(x)_{10} | 1000 \leq x \leq 9999\}$. Let A_i be the set of numbers in x whose ith digit is i. Compute $N(A_1 \cup A_2 \cup A_3 \cup A_4)$.

$$N(A_1 \cup A_2 \cup A_3 \cup A_4) = N(A_1) + N(A_2) + N(A_3) + N(A_4)$$
$$- N(A_1 \cap A_2) - N(A_1 \cap A_3) - N(A_1 \cap A_4)$$
$$- N(A_2 \cap A_3) - N(A_2 \cap A_4) - N(A_3 \cap A_4)$$
$$+ N(A_1 \cap A_2 \cap A_3) + N(A_1 \cap A_2 \cap A_4)$$
$$+ N(A_1 \cap A_3 \cap A_4) + N(A_2 \cap A_3 \cap A_4)$$
$$- N(A_1 \cap A_2 \cap A_3 \cap A_4)$$
$$= 1000 + 900 + 900 + 900 - 100 - 100 - 100$$
$$- 90 - 90 - 90 + 10 + 10 + 10 + 9 - 1$$
$$= 3168 \quad \#$$

K. ORDERED SETS

We now wish to introduce some generalizations of the notion of order as conveyed by a sequence. Relations having the generalized ordering properties are known as *ordering relations*.

Definition

A relation $R \subseteq A \times A$ is a preorder if R is reflexive and transitive.

Definition

A relation $R \subseteq A \times A$ is a partial order if R is reflexive, antisymmetric, and transitive. The set A is said to be partially ordered by R. Whenever aRb or bRa, a is said to be comparable to b; otherwise, they are incomparable.

Example

The set of natural numbers is partially ordered by the relation $R(\leqq)$. We can check that $R(\leqq)$ is:

(a) Reflexive: for any a, $a \leqq a$.
(b) Antisymmetric: for any a, b, if $a \leqq b$ and $b \leqq a$, then $a = b$.
(c) Transitive: if $a \leqq b$ and $b \leqq c$, then $a \leqq c$.

Example

A sequence $(a_0, a_1, \ldots a_r)$ is said to be a prefix of another sequence $(b_0, b_1, \ldots, b_s)$ if $a_0 = b_0$, $a_1 = b_1, \ldots$, and $a_r = b_r$. The set of sequences over a set A is partially ordered by the relation R (*is a prefix of*).

(a) Reflexive: for any sequence s, s is a prefix of s.
(b) Antisymmetric: for any sequences s_1 and s_2, if s_1 is a prefix of s_2 and s_2 is a prefix of s_1, they must have the same number of elements and are equal term by term. Hence $s_1 = s_2$.
(c) Transitive: if a is a prefix of b and b is a prefix of c, then a is a prefix of c.

Let $A =$ the set of sequences over $\{a, b, c\}$ and $P =$ "is a prefix of"; then

$$(a, b) \; P \; (a, b, a)$$

$$(a, b) \; P \; (a, b, c)$$

while (a, b, a) and (a, b, c) are incomparable.   #

A Hasse diagram is a pictorial representation of a finite partial order. In the representation, the elements are shown as vertices and, if $a \geqq b$, the vertex a is shown above the vertex b. Furthermore, there is a line connecting two related vertices if they are related, and the relationship cannot be deduced from the reflexive and transitive properties of partial orders.

Example

$$\{(a, a), (a, b), (a, c), (b, b), (c, c)\} = R_1.$$

Hasse diagram of R_1

Example

The partial order "is a prefix of" on the set of elements $\{(1,1),\ (1,1,2),\ (1,1,3),\ (1,1,3,1),\ (1,2),\ (1,2,1),\ (1,2,2),\ (1,2,2,1)\}$.

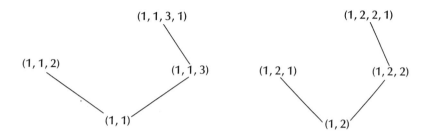

Definition

A relation *R* is said to be connected *if, for every pair of different elements a and b, either aRb or bRa.*

Definition

A linear ordering *is a partial ordering that is connected.*

Example

Let *A* be a linearly ordered finite alphabet. Define an ordering on words over *A* by comparing pairs of different words. Suppose $s = s_0 s_1 \ldots s_m$ and $t = t_0 t_1 \ldots t_n$; then we define $R(\leq)$ as follows.

$s \leq t$ if $s_i = t_i$ for $i = 0, 1, 2, \ldots, r$ and $s_{r+1} < t_{r+1}$

or

s is a prefix of t.

Consider the words "apple," "apply," and "application," with the usual ordering on the alphabet. From the definition of $R(\leq)$, we can see that

$$\text{apple} < \text{application} < \text{apply}$$

This linear ordering is known as *lexicographic* ordering. #

Definition

A set with an ordering relation is well-ordered *if every nonnull subset has a least element.*

Examples

The natural numbers are well-ordered. The integers are not well-ordered. There is no smallest negative number.   #

The property of well-ordering of the natural numbers (and its generalizations for partially ordered sets) is often used to prove that a program with an iterative control structure will terminate after a finite number of iterations.

Example

The following program computes the greatest common divisor of a pair of positive integers.

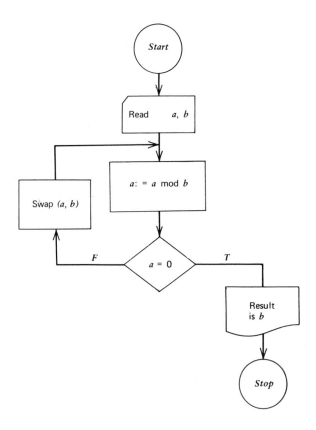

The program terminates, since $a + b$ is a positive integer whose value is decreased each time around the loop.

The following program to check for a pair of positive integers, a, b, if b divides a is incorrect; if b does not divide a, the program never terminates.

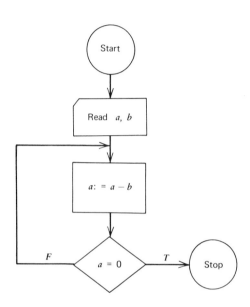

Note that the distinction between these two programs is that in this second program $a + b$ is not always positive, so well-ordering cannot apply.   #

L. OPERATIONS ON RELATIONS

We can define operations on relations that allow us to generate new relations from given ones. First we have the following unary operations.

Definition

The converse of a relation R is the relation $\check{R}$ given by $\check{R} = \{(x, y)|\ (y, x)\ is\ in\ R\}$. Another notation for $\check{R}$ is R^{-1}. (Note that the domain of $\check{R}$ is taken as the range of R, and the range of $\check{R}$ is taken to be the domain of R.)

Example

$R(<) = \{(1, 2),\ (2, 3),\ (1, 3)\}$ and $R(>) = \check{R}(<) = \{(2, 1),\ (3, 2),\ (3, 1)\}$.   #

Fact.   If $S = \check{R}$, then $R = \check{S}$.

Definition

If R is a given relation, defined as a subset of $A \times B$, the complement of R (with respect to $A \times B$), denoted $\bar{R}$, is given by $\bar{R} = \{(x, y)|\ (x, y)$ is in $A \times B$ and (x, y) is not in $R\} = (A \times B) - R$.

Fact.   If $S = \bar{R}$, then $R = \bar{S}$.

Example

If $R \leqq \{1, 2, 3\} \times \{1, 2, 3, 4\}$ is $R = \{(1, 2),\ (2, 3),\ (1, 3)\}$, then $\bar{R} = \{(1, 1), (1, 4), (2, 1), (2, 2), (2, 4), (3, 1), (3, 2), (3, 3), (3, 4)\}$.   #

Next we have binary operations on relations. The definition of union and intersection of a pair of relations (defined as subsets of the same Cartesian product) is clear, since relations are sets, and the union and intersection are just set union and set intersection, respectively.

Definition

The composition of two relations R_1, R_2, where $R_1 \subseteq A \times B$ and $R_2 \subseteq B \times C$, is the relation $R_1 \circ R_2 = \{(x, y)|\ for\ some\ z,\ (x, z)$ is in R_1, and (z, y) is in $R_2\}$.

Example

$R_1 = \{(a, b),\ (a, c)\},\ R_2 = \{(b, d),\ (c, e),\ (c, f)\}$, and $R_1 \circ R_2 = \{(a, d),\ (a, e),\ (a, f)\}$.   #

Using the alternate notation for ordered pairs, makes the composition clearer. Recall that (a, b) may also be denoted $\dfrac{a}{b}$. Then we may define the composition of two relations R_1 and R_2 as the relation consisting of all products of the form $\dfrac{a}{b} \cdot \dfrac{b}{c}$, where $\dfrac{a}{b}$ is in R_1 and $\dfrac{b}{c}$ is in R_2 and where we define $\dfrac{a}{b} \cdot \dfrac{c}{d}$ as (an apparent) cancellation.

Example

$R_1 = \left\{\dfrac{a}{b}, \dfrac{a}{c}\right\}, R_2 = \left\{\dfrac{b}{d}, \dfrac{c}{e}, \dfrac{c}{f}\right\}$, and $R_1 \circ R_2 = \left\{\dfrac{a}{d}, \dfrac{a}{e}, \dfrac{a}{f}\right\}$.   #

Fact.   Composition of relation is associative; that is, $(R_1 \circ R_2) \circ R_3 = R_1 \circ (R_2 \circ R_3)$.

Example

Let $R_1 = \left\{\dfrac{a}{b}, \dfrac{a}{c}\right\}$, $R_2 = \left\{\dfrac{b}{d}, \dfrac{c}{e}, \dfrac{c}{f}\right\}$, and $R_3 = \left\{\dfrac{d}{a}, \dfrac{f}{b}, \dfrac{f}{c}\right\}$. Then

$(R_1 \circ R_2) \circ R_3 = \left\{\dfrac{a}{d}, \dfrac{a}{e}, \dfrac{a}{f}\right\} \circ \left\{\dfrac{d}{a}, \dfrac{f}{b}, \dfrac{f}{c}\right\} = \left\{\dfrac{a}{a}, \dfrac{a}{b}, \dfrac{a}{c}\right\}$, while $R_1 \circ (R_2 \circ R_3) =$

$\left\{\dfrac{a}{b}, \dfrac{a}{c}\right\} \circ \left\{\dfrac{b}{a}, \dfrac{c}{b}, \dfrac{c}{c}\right\} = \left\{\dfrac{a}{a}, \dfrac{a}{b}, \dfrac{a}{c}\right\}$. Thus $(R_1 \circ R_2) \circ R_3 = R_1 \circ (R_2 \circ R_3)$
and may be denoted $R_1 \circ R_2 \circ R_3$.   #

Let E denote the identity relation (the universe being understood) $E = \{(x, x) \mid \text{every } x \text{ in } I\}$. Then an equivalence relation may be concisely defined in terms of the operations on relations as follows:

$$E \subseteq R \qquad (R \text{ is reflexive}) \qquad (1\text{-}8)$$

$$R = \breve{R} \qquad (R \text{ is symmetric}) \qquad (1\text{-}9)$$

$$R \circ R \subseteq R \qquad (R \text{ is transitive}) \qquad (1\text{-}10)$$

We may, in fact, recast all the previous definitions of set properties as follows.

Property of R	Defining Equation
Reflexivity	$E \subseteq R$
Irreflexivity	$R \cap E = \emptyset$
Symmetry	$R = \breve{R}$
Antisymmetry	$R \cap \breve{R} \subseteq E$
Asymmetry	$(R \cap \breve{R}) = \emptyset$
Transitivity	$R \circ R \subseteq R$

M. FUNCTIONS

The idea of a function is to associate a unique value (or result) with a given argument. The concept of functionality is basic to the idea of (deterministic) computation. A function is visualized as a rule that has exactly one value in the range corresponding to an element in the domain. We will also extend the notion of functionality to *partial functions*; a partial function has at most one value in the range corresponding to an element in the domain, but possibly no values for some arguments. Partial functions are used as theoretical models of (deterministic) computation, which may loop for some inputs and thus return no value.

Definition

A relation $R \subseteq A \times B$ is a partial function from A to B if there are no two ordered pairs in R having the same first element but a different second element; that is, if (x, y) is in R and (x, z) is in R, then y = z. We sometimes say that, in this case, R is a functional relation from A to B.

Examples

$R_1 = \{(a, b), (a, c)\}$ is not a function. $R_2 = \breve{R}_1 = \{(b, a), (c, a)\}$ is a function.   #

Notation. We will use the letter F when we want to denote a (partial) function specifically. If (x, y) is in F, then we write xFy, Fx = y, or F(x) = y. (Some authors also use xF = y.)

A function is called *total* if $F \subseteq A \times B$ and Fa is defined for each a in A. A total function is also called a *mapping*.

Fact. Composition of total functions yields a total function. Composition of partial functions yields a partial function.

Notation. If F is a functional relation from K_1 to K_2, we write $F: K_1 \to K_2$; for $k_1 F k_2$, we write $F: k_1 \mapsto k_2$.

Definition

Let F be a function and x an element of the range of F; then $\{y \mid yFx\}$ is called the set of preimages of x.

Definition

A function is $1-1$ if its converse is functional.

Note that since functions are relations, converses are defined. Alternatively, we could define a $1-1$ function as a function in which the set of preimages of an element x is a unit set containing exactly one element.

Definition

A function $F \subseteq A \times B$ is onto if every element in B has a preimage in A.

Notation. One-to-one functions are also called *injections*; onto functions are also called *surjections*; and functions that are both $1-1$ and onto are called *bijections*.

Clearly a necessary condition for two sets A and B to be equal is that there be a bijection between A and B.

Fact.   There is no bijection between a set S and its power set $P(S)$.

Proof.   First, for finite sets this is obvious by counting, so our proof must take infinite sets into account; it will be a "diagonalization." We will assume that bijection between S and $P(S)$ exists and deduce a contradiction.

By assumption there is a bijection $x \leftrightarrow S_x$ taking elements of S into subsets of S and vice versa. Some elements are members of the subsets into which they are mapped; call these type 1 elements. The elements that are not members of the subsets into which they are mapped will be called type 2 elements. Now consider the set S_z of all type 2 elements and the element z to which it is mapped: $S_z = \{x \mid x \notin S_x\}$.

Suppose first that $z \in S_z$. Then, by definition of S_z, since S_z consists of all type 2 elements, z must be of type 2. But if z is of type 2, by the definition of type 2, $z \notin S_z$. Thus $z \in S_z$ implies $z \notin S_z$.

Conversely, suppose $z \notin S_z$. Then, by definition of the type 2 element, z is a type 2 element. But then, by the definition of S_z, $z \in S_z$. Thus $z \notin S_z$ implies $z \in S_z$.

Therefore, $z \in S_z$ if and only if $z \notin S_z$.

(Note that because we did not assume that S is enumerable, we cannot construct a diagonal visualization. However, since the type of argument is similar, this is still called a proof by diagonalization.) #

Definition
     A permutation is a function from a finite set onto itself.

Example
     $f = \{(a, b), (b, a), (c, c)\}$ is a permutation of $\{a, b, c\}$.   #

Fact.   Every permutation is also a $1-1$ function. Hence the converse of a permutation is a permutation.

N. MATRICES

The operations on relations and functions are often conveniently organized as computations on sets of matrices. Each matrix is taken to represent a relation, and multiplication of matrices (under the appropriate algebraic rules) is a convenient way to compute relational composition. First, the algebraic operations $\vee$ and $\wedge$ are introduced and the application of these operations in matrix computations is defined. This is done in such a way that the "usual" algebraic operations on matrices, $+$ and $\times$, and those needed in relational computations are both seen to be special cases of the more general pattern.

Let θ be any binary operation defined as a total function, $\theta: A^2 \to A$, where A is the set over which θ is defined. We extend the definition of θ to a mapping from A^k to A for each $k \geq 1$ as follows.[4]

$$\theta: a_0 \mapsto a_0$$

$$\theta: (a_0, a_1, \ldots, a_n) \mapsto a_0 \theta(\theta(a_1, \ldots, a_n))$$

Examples

Let us define the following binary operations from $\{0, 1\}^2 \to \{0, 1\}$:

$\wedge$	0	1
0	0	0
1	0	1

$\vee$	0	1
0	0	1
1	1	1

and the unary operation:

$\sim$	
0	1
1	0

These operations may be shown in relational form as follows.

$$\wedge = \{((0, 0), 0), ((0, 1), 0), ((1, 0), 0), ((1, 1), 1)\}$$
$$\vee = \{((0, 0), 0), ((0, 1), 1), ((1, 0), 1), ((1, 1), 1)\}$$
$$\sim = \{(0, 1), (1, 0)\}$$

Applying our recursive definition of $\vee$,

$$\vee(0, 1, 0, 1) = 0 \vee (\vee(1, 0, 1))$$
$$\vee(1, 0, 1) = 1 \vee (\vee(0, 1))$$
$$\vee(0, 1) = 0 \vee (\vee(1))$$
$$\vee(1) = 1$$

[4]We have used here the APL operations of "compression" to extend a binary operation to a unary operation on sequences.

Thus

$$v(0, 1) = 0 \vee 1 = 1$$
$$v(1, 0, 1) = 1 \vee 1 = 1$$
$$v(0, 1, 0, 1) = 0 \vee 1 = 1 \qquad \#$$

We may also extend θ to be a binary operation on sequences of equal length, as follows.

$$\theta : ((r_0, r_1, \ldots, r_k), (s_0, s_1, \ldots, s_k)) \to (r_0 \theta s_0, r_1 \theta s_1, \ldots, r_k \theta s_k)$$

Example

Let $+$ be the usual arithmetic operation; then

$$(1, 2, 3, 5) + (2, 4, 6, 8) = (3, 6, 9, 13) \quad \#$$

Definition

An $m \times n$ matrix *is an arrangement of m rows and n columns of elements.*

Example

$$2 \times 2 : \begin{pmatrix} a & b \\ c & d \end{pmatrix} \qquad 3 \times 4 : \begin{pmatrix} 1 & 2 & 3 & 4 \\ a & b & c & d \\ e & f & 1 & \emptyset \end{pmatrix} \quad \#$$

Notation. An $r \times s$ matrix, M, will be denoted $\mathbf{M}$; $\mathbf{M}_{ij}$ denotes the element in row i, column j, $\mathbf{M}_i$ denotes the ith row, $\mathbf{M}_i = (M_{i0}, M_{i1}, \ldots, M_{i,s-1})$, $\mathbf{M}^j$ denotes the (transpose of the) jth column, and $\mathbf{M}^j = (M_{0j}, M_{ij}, \ldots, M_{r-1,j})$.

Definition

The *transpose of an* $n \times m$ *matrix A is an* $m \times n$ *matrix* $\mathbf{A}^T$, *where* $A_{ij}^T = A_{ji}$. *A row may be considered as a* $1 \times n$ *matrix, and a column may be thought of as an* $n \times 1$ *matrix. Thus the transpose of a row,* $\mathbf{M}_i$, *is a column,* $\mathbf{M}_i^T$, *and vice versa.*

Let M be an $r \times s$ matrix. Then M can also be written as a sequence of sequences (where for convenience of visualization we can write some sequences vertically and some horizontally).

Notation.   The vertical sequence $\mathbf{V} = \begin{pmatrix} a_0 \\ \vdots \\ a_r \end{pmatrix}$ is also written as $\mathbf{U}^T$, where $U = (a_0, \ldots, a_r)$. Then $M = \begin{pmatrix} M_0 \\ \vdots \\ M_{r-1} \end{pmatrix}$, where $M_i =$ $(M_{i0}, M_{i1}, \ldots, M_{i,s-1})$. Similarly, $M = (M^{0T}, M^{1T}, \ldots, M^{(s-1)T})$, where $M^i = (M_{0j}, M_{1j}, \ldots, M_{r-1,j})$.

Definition

Let M be an $r \times s$ matrix and N be an $s \times t$ matrix. The generalized product $M \, {}^{\theta_1}_{\theta_2} N$ is an $r \times t$ matrix P whose (i, j)th element is given by $P_{ij} = \theta_1(M_i \theta_2 N^j)$. (Note. The number of columns in the first matrix in the product, M, must equal the number of rows of the second matrix of the product, N.)

Example

Usual matrix multiplication is defined as $C = A \, {}^{+}_{\times} B$. Let $A =$ $\begin{pmatrix} 1 & 2 & 3 \\ 4 & 5 & 6 \end{pmatrix}$ and $B = \begin{pmatrix} 1 & 2 & 3 & 4 \\ 5 & 6 & 7 & 8 \\ 9 & 10 & 11 & 12 \end{pmatrix}$. Then

$A_1 = (1, 2, 3)$	$B^1 = (1, 5, 9)$
$A_2 = (4, 5, 6)$	$B^2 = (2, 6, 10)$
	$B^3 = (3, 7, 11)$
	$B^4 = (4, 8, 12)$

$A_1 \times B^1 = (1, 10, 27)$	$A_2 \times B^1 = (4, 25, 54)$
$A_1 \times B^2 = (2, 12, 30)$	$A_2 \times B^2 = (4, 30, 60)$
$A_1 \times B^3 = (3, 14, 33)$	$A_2 \times B^3 = (12, 35, 66)$
$A_1 \times B^4 = (4, 16, 36)$	$A_2 \times B^4 = (16, 40, 72)$

$$C_{ij} = + (A_i \times B^j)$$

$C_{11} = 38$	$C_{21} = 83$
$C_{12} = 44$	$C_{22} = 94$
$C_{13} = 50$	$C_{23} = 113$
$C_{14} = 56$	$C_{24} = 128$

$$C = \begin{pmatrix} 38 & 44 & 50 & 56 \\ 83 & 94 & 113 & 128 \end{pmatrix} \quad \#$$

O. MATRIX REPRESENTATION OF RELATIONS

Suppose that $R \subseteq M \times N$ is a relation and M and N are both finite sets; then R may be represented as a matrix, M_R.

First observe that if S is any finite set, it is enumerable, so there is an indexed set whose elements are the elements of S. So M and N can each be indexed.

Example
    If $M = \{a, b, c\}$ and $N = \{1, 2, 3, 4\}$, we might choose

$$F: a \mapsto 0, \qquad b \mapsto 1, \qquad c \mapsto 2$$
$$G: 1 \mapsto 0, \qquad 2 \mapsto 1, \qquad 3 \mapsto 2, \qquad 4 \mapsto 3$$

Then **M** is a sequence such that set $(\mathbf{M}) = M$ and $M_0 = a$, $M_1 = b$, $M_2 = c$. The same applies for **N**. Now R may be represented as an $|\mathbf{M}| \times |\mathbf{N}|$ matrix, $\mathbf{M}_R$, where $|\mathbf{M}|$ denotes the cardinality of **M**. If **M** is finite, then the cardinality of **M** is the number of elements of **M**. In $\mathbf{M}_R$, the matrix of the relation R, the element $M_{ij} = 1$ if and only if (M_i, N_j) is in R.

Example
    $R(\leqq) \subseteqq \{1, 2, 3, 4\} \times \{1, 2, 3, 4\}$

    $R(\leqq) = \{(1, 1), \ (1, 2), \ (1, 3), \ (1, 4), \ (2, 2), \ (2, 3), \ (2, 4), \ (3, 3),$ $(3, 4), \ (4, 4)\}$

Using the "natural" indexing, that is, $1 \mapsto 0, 2 \mapsto 1, 3 \mapsto 2, 4 \mapsto 3$, $\mathbf{M}_R$ is a 4×4 matrix of 1's and 0's.

$$\mathbf{M}_R = \begin{pmatrix} 1 & 1 & 1 & 1 \\ 0 & 1 & 1 & 1 \\ 0 & 0 & 1 & 1 \\ 0 & 0 & 0 & 1 \end{pmatrix} \qquad \mathbf{M}_{\check{R}} = \begin{pmatrix} 1 & 0 & 0 & 0 \\ 1 & 1 & 0 & 0 \\ 1 & 1 & 1 & 0 \\ 1 & 1 & 1 & 1 \end{pmatrix} \qquad \#$$

If $R_1 \subseteqq A \times B$ and $R_2 \subseteqq B \times C$ and, in $\mathbf{M}_{R_1}$ and $\mathbf{M}_{R_2}$ the same indexing is used for the set B, the matrix of $R_1 \circ R_2$, $\mathbf{M}_{R_1 \circ R_2}$, is calculated by

$$\mathbf{M}_{R_1 \circ R_2} = \mathbf{M}_{R_1} \overset{\vee}{\underset{\wedge}{}} \mathbf{M}_{R_2}$$

Example
    Let $R_1 \subseteqq \{a, b, c\} \times \{l, m, n, o\}$ and $R_1 = \{(a, l), (a, m), (b, m),$ $(b, n), (c, l), (c, o)\}$ with indexing $a \mapsto 0$, $b \mapsto 1$, $c \mapsto 2$ and $l \mapsto 0$, $m \mapsto l$, $n \mapsto 2$, $o \mapsto 3$; then

$$\mathbf{M}_{R_1} = \begin{pmatrix} 1 & 1 & 0 & 0 \\ 0 & 1 & 1 & 0 \\ 1 & 0 & 0 & 1 \end{pmatrix}$$

Let $R_2 \subseteq \{l, m, n, o\} \times \{u, v, w\}$ and $R_2 = \{(l, u), (m, u), (m, v), (m, w), (n, u), (n, v), (o, w)\}$. Using the same indexing for l, m, n, o and $u \to 0$, $v \to 1$, $w \to 2$,

$$M_{R_2} = \begin{pmatrix} 1 & 0 & 0 \\ 1 & 1 & 1 \\ 1 & 1 & 0 \\ 0 & 0 & 1 \end{pmatrix}$$

Then

$$M_{R_1} \overset{\vee}{\underset{\wedge}{}} M_{R_2} = \begin{pmatrix} 1 & 1 & 1 \\ 1 & 1 & 1 \\ 1 & 0 & 1 \end{pmatrix}$$

Also, by direct calculation,

$$R_1 \circ R_2 = \left\{ \frac{a}{l}, \frac{a}{m}, \frac{b}{m}, \frac{b}{n}, \frac{c}{l}, \frac{c}{o} \right\} \circ \left\{ \frac{l}{u}, \frac{m}{u}, \frac{m}{v}, \frac{m}{w}, \frac{n}{u}, \frac{n}{v}, \frac{o}{w} \right\}$$

$$= \left\{ \frac{a}{u}, \frac{a}{v}, \frac{a}{w}, \frac{b}{u}, \frac{b}{v}, \frac{b}{w}, \frac{c}{u}, \frac{c}{w} \right\}$$

and, with the same indexing as previously,

$$M_{R_1 \circ R_2} = \begin{pmatrix} 1 & 1 & 1 \\ 1 & 1 & 1 \\ 1 & 0 & 1 \end{pmatrix} \quad \#$$

P. SOME SPECIAL MATRICES

The matrix of a function has at most "1" per row; the remaining elements are "0," since the matrix of any relation consists solely of 0's and 1's. If, in addition, the matrix of a function has only one 1 per column, the function is one to one. Recall that a finite one-to-one function is called a *permutation*.

The matrix of an equivalence relation is a square matrix whose rows can be permuted so that, with the same indexing on rows and columns, the following type of matrix results.

$$\begin{array}{c|c} U & \emptyset \\ \hline 0 & \alpha \end{array}$$

where U is a square matrix with $U_{ij} = 1$ for all i, j (U is the matrix of the universal relation); $\emptyset$ is a matrix of all 0's ($\emptyset$ is the matrix of an empty relation); and α is the matrix of an equivalence relation.

It is both a necessary and sufficient condition for R to be an equivalence relation that its matrix can be rearranged in this form.

Example

Let $R_p \subseteq \{1, 2, 3, 4, 5\} \times \{1, 2, 3, 4, 5\}$ denote the equivalence relation in which the even integers are one equivalence class and the odd integers are another. The matrix of R_p, M_{R_p} is

	1	2	3	4	5
1	1	0	1	0	1
2	0	1	0	1	0
3	1	0	1	0	1
4	0	1	0	1	0
5	1	0	1	0	1

Rearranging, this matrix becomes

	1	3	5	2	4
1	1	1	1	0	0
3	1	1	1	0	0
5	1	1	1	0	0
2	0	0	0	1	1
4	0	0	0	1	1

#

Q. TRANSITIVE CLOSURE

Suppose that P is any property of relations. The P-closure of a relation R is the smallest relation R', such that $R \subseteq R'$ and R' has property P.

Examples

The reflexive closure of $R(<)$ is $R(\leq)$. Let $R \subseteq \{1, 2, 3\} \times \{1, 2, 3\}$ and $R = \{(1, 2), (1, 3)\}$. The reflexive closure of R is $\{(1, 1), (1, 2), (1, 3), (2, 2), (3, 3)\}$. The symmetric closure of R is $\{(1, 2), (2, 1), (1, 3), (3, 1)\}$. The reflexive and symmetric closure of R is $\{(1, 1), (1, 2), (1, 3), (2, 1), (2, 2), (3, 1), (3, 3)\}$.   #

Note that the reflexive and symmetric closure is both *the reflexive closure of the symmetric closure* and *the symmetric closure of the reflexive closure*. Let R be the relation in question. The reflexive closure of R is $R \cup E$, where E is the identity relation. The symmetric closure of R is $R \cup \breve{R}$. The symmetric closure of the reflexive closure is:

$$(R \cup E) \cup (R \cup E)$$
$$= (R \cup E) \cup (\breve{R} \cup \breve{E}) \qquad \text{property of converse}$$
$$= (R \cup E) \cup (\breve{R} \cup E) \qquad E = \breve{E} \text{ (i.e., identity is symmetric)}$$
$$= (R \cup \breve{R}) \cup E \qquad \text{commutativity and associativity and}$$
$$\qquad\qquad\qquad\qquad E \cup E = E$$

But this last equation is just the definition of the reflexive closure of the symmetric closure of R.

The transitive closure of R is the smallest relation containing R that is transitive. Let $R^+ = R \cup R^2 \cup R^3 \cup \cdots$ where $R^{k+1} = R^k \circ R$ for $k \geqq 1$.

Fact. R^+ is the transitive closure of R.

Proof.   If R is transitive, the transitive closure of R is R itself. But if R is transitive, $R^2 \subseteqq R$, and $R^k \subseteqq R$ for every $k \geqq 1$, as can be readily proved by induction. Thus, if R is transitive, $R^+ = R$.

If R is not transitive, we claim that R^+ is transitive, since $(R^+)^+$ contains only integral powers of R and so $(R^+)^+ \subseteqq R^+$. But since $(R^+)^+$ has a term $R^+ \circ R^+$, we have that $R^+ \circ R^+ \subseteq R^+$, showing that R^+ is, indeed, transitive.

Furthermore, if S is any transitive relation on the same set as R and $R \subseteq S$, then $R^+ \subseteq S^+$, since it is true for each k that $R^k \subseteq S^k$. But $S^+ = S$, so $R^+ \subseteq S$.

Since R^+ is transitive and is contained in any transitive relation containing R, R^+ is the smallest transitive relation containing R, and so is the transitive closure of R.   #

Fact. R^+ is the smallest transitive relation containing R. Note that if $R \subseteq A \times A$ and if A is finite, then R is finite; $|R| \subseteq |A|^2$.

Fact.   If R is finite and $|A| = k$, then $R^k = R^{k+1}$, so that to compute R^+ over a finite set requires only a finite computation. (This fact follows readily from the graph interpretation of a relation, given in the next chapter.)

Examples

(a) $R = \left\{ \dfrac{a}{a}, \dfrac{a}{b} \right\}$, $R^2 = \left\{ \dfrac{a}{a}, \dfrac{a}{b} \right\}$, and $R^+ = R$.

In fact, noting that R is transitive, we may observe directly that $R^+ = R$, since it is the smallest transitive relation containing R.

(b) $R = \left\{ \frac{a}{b}, \frac{b}{c}, \frac{c}{d} \right\}$, $R^2 = \left\{ \frac{a}{c}, \frac{b}{d} \right\}$, $R^3 = \left\{ \frac{a}{d} \right\}$, and

$$R^+ = \left\{ \frac{a}{b}, \frac{b}{c}, \frac{c}{d}, \frac{a}{c}, \frac{b}{d}, \frac{a}{d} \right\}. \quad \#$$

We have seen that R is transitive if and only if $R \circ R \subseteq R$. Thus, if R is transitive, $R^k \subseteq R$ for every k, and thus $R^+ = R$. Even if R is not transitive, R^+ is transitive and, consequently, $(R^+)^+ = R^+$. Moreover, if S is any transitive relation on A, $S \subseteq A \times A$ and $R \subseteq S$; then $R^+ \subseteq S$.

Fact.   The reflexive closure of the transitive closure of R is equal to the transitive closure of the reflexive closure of R. Hence, it is often referred to as the reflexive-transitive closure of R.

R. PARTIALLY ORDERED SETS AND LATTICES

A *poset* (*partially ordered set*) is a set with a partial ordering.

Definition

Let $(P, \leqq)$ be any poset and let a, b in P be arbitrary elements. Then an element d in P is called the greatest lower bound (*glb*) of a and b when $d \leqq a$ and $d \leqq b$ implies that for any x, such that $x \leqq a$ and $x \leqq b$, $x \leqq d$.

We represent posets diagramatically by showing elements connected by straight lines, as in Figure 1-4. If the point labeled a is connected to the point labeled b, and if a is higher in the diagram than b (i.e., closer to the top of the page), then $a \geqq b$. Also, to avoid cluttering the diagram, we make use of transitivity

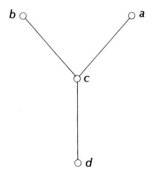

Figure 1-4.   Poset.

by not showing the "lines" of the diagram that could be inferred from transitivity.

Example

The relation, R, represented by Figure 1-4 is $(S \cup E)^+$, where
$S = \left\{\dfrac{d}{c'}, \dfrac{c}{b'}, \dfrac{c}{a}\right\}$.   #

We introduce now the notion of complementarity for set expressions. $\bar{A}$ is, of course, the complement of A. The universal set, U, is the complement of $\emptyset$. Finally, the "complement" of intersection is union.

Principle of Complementarity
The complement of any set S denoted by an expression E is denoted by $\bar{E}$ obtained from E by interchanging $(\cap, \cup)$, $(U, \emptyset)$, and $(A, \bar{A})$ for every A.

Example

$S = A \cup (\bar{B} \cap (C \cup \bar{D}))$ and $\bar{S} = \bar{A} \cap (B \cup (\bar{C} \cap D))$.   #

Note that if $T = A \cup B$, $\bar{T} = \bar{A} \cap \bar{B}$. Using these equations, we have $A \cup B = \overline{(\bar{A} \cap \bar{B})}$ and $A \cap B = \overline{(\bar{A} \cup \bar{B})}$. These equations are DeMorgan's laws (Section E).

Note that if $A = B$, $\bar{A} = \bar{B}$. Furthermore, if $A \subseteq B$, $\bar{B} \subseteq \bar{A}$.

The pairs $(U, \emptyset)$, $(\cap, \cup)$, and $(\subseteq, \supseteq)$ are called dual pairs. The principle of duality follows from the principle of complementarity.

Principle of Duality
Any statement involving arbitrary sets is true if and only if its dual is true (where the dual of a statement is obtained by replacing each element by its dual).

Examples

$$A \cap \bar{A} = \emptyset \qquad\qquad A \cup \bar{A} = U$$
$$A \cap B \subseteq A \qquad\qquad A \cup B \supseteq A$$
$$A \cup (B \cap C) \subseteq A \cup B \qquad A \cap (B \cup C) \supseteq A \cap B \quad \#$$

Definition

The least upper bound (*lub*) is the dual concept of the glb. Let P be a poset. An element d in P is the lub of a and b when *d ≧ a and d ≧ b implies that for any x, such that x ≧ a and x ≧ b, then x ≧ d.*

Note that if a glb exists, it is unique. Similarly, if a lub exists, *it* is unique.

Notation.   The lub of x and y is denoted $x \vee y$. The glb of x and y is denoted $x \wedge y$.

Definition

    A lattice *is a poset in which any two elements have a glb and a lub.*

Examples

    Figure 1-5a, 1-5b, and 1-5c shows a number of lattices. Also shown is one poset that fails to be a lattice (Figure 1-5d) because one can find pairs of elements without a glb or lub.   #

Exercise.   Construct a poset in which every pair of elements has a lub, but at least one pair of elements has no glb. Does your poset have the smallest number of elements that can fail to satisfy the definition of a lattice?

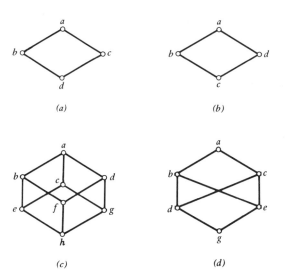

Figure 1-5.   Examples. (a) A lattice. $a \wedge a = a$, $a \wedge b = b$, $a \wedge c = c$, $a \wedge d = d$, $b \wedge c = d$, $b \wedge b = b$, $d \wedge b = d$, $c \wedge d = d$. (b) A lattice. (c) A lattice. (d) Not a lattice. Observe that b and c have no glb because $e \leqq b$, $e \leqq c$ and $d \leqq b$, $d \leqq c$, but d and e are incomparable.

S. VIEWING RELATIONS AS FUNCTIONS

A relation R from X to Y (i.e., $R \subseteq X \times Y$) is a function $f: P(X) \to P(Y)$, that is, a function mapping the power set of X to the power set of Y that satisfies the equations $f(\cup A_i) = \cup f(A_i)$ where $\{A_i | i \text{ is in } I\}$ is a family of subsets of X indexed by some set I (and where $\cup$ stands for union).

f may be constructed as follows: $\emptyset \to \emptyset$. Then, for each element a, $\{a\} \to \{x | (a, x) \text{ is in } R\}$, and for each m elements $a_1, a_2, \ldots, a_m$, $\{a_1, a_2, \ldots, a_m\} \to \{x | (a_i, x) \text{ is in } R \text{ for some } 1 \le i \le m\}$.

Example

$$R = \left\{ \frac{a}{b}, \frac{a}{c}, \frac{b}{c}, \frac{b}{d}, \frac{c}{c}, \frac{c}{e} \right\} \quad X = \{a, b, c\} \quad Y = \{b, c, d, e\}$$

$P(X) = \{\emptyset, \{a\}, \{b\}, \{c\}, \{a, b\}, \{a, c\}, \{b, c\}, \{a, b, c\}\} \quad f: \emptyset \mapsto \emptyset$

$\{a\} \mapsto \{b, c\}$ $\{a, b\} \mapsto \{b, c, d\}$ $\{a, b, c\} \mapsto \{b, c, d, e\}$

$\{b\} \mapsto \{c, d\}$ $\{a, c\} \mapsto \{b, c, e\}$

$\{c\} \mapsto \{c, e\}$ $\{b, c\} \mapsto \{c, d, e\}$ #

T. ANOTHER VIEW OF FUNCTIONS

We have seen that a binary function can be viewed as a mapping from a Cartesian product to a set. In this view, $F = \{((a, b), c)\}$ and, when $((a, b), c)$ is in F, we also write $F(a, b) = c$ or $aFb = c$. An alternative description of the same mathematical object is to parametrize it as a set of unary functions, $\{F_a\}$, one such function for each element a in A, as follows: $F_a(b) = F(a, b) = c$. Likewise, we could parametrize on the second element to obtain $G_b(a) = F(a, b) = c$.

Example

$\cap: \{0, 1\} \times \{0, 1\} \to \{0, 1\}$ is   defined   by $\cap = \{((0, 0), 0),$ $((0, 1), 0), ((1, 0), 0), ((1, 1), 1)\}$ or, equivalently, in tabular form

$\cap$	0	1
0	0	0
1	0	1

A parametric description is

$\cap_0 = \{(0, 0), (1, 0)\}$ $\cap_1 = \{(0, 0), (1, 1)\}$ #

The parametric description is useful in describing finite state machines.

Finite state machines have many applications. Perhaps the most familiar one is as lexical scanner in the initial phase of a compiler. The purpose of the lexical scanner is to assemble subsequences of input symbols into *tokens*, which are entered into the symbol table of the compiler. Lexical scanners are typically finite state machines.

Definition

A finite state machine, M, is a mathematical model consisting of a finite set of states, S, a finite set of symbols, Σ, a next state function, $\sigma: S \times \Sigma \to S$, an output function, $\beta: S \to \{0, 1\}$, and a starting state, s_0 in S.

The machine starts in state s_0 and a sequence of symbols from Σ is applied to the machine. When the machine is in state s_i and symbol σ_j is applied, the new state of the machine s'_i is determined by the next state function, $s'_i = \delta(s_i, \sigma_j)$. When the complete sequence has been applied, the machine is in some state, s, and if $\beta(s) = 1$, we say that the sequence of symbols is accepted; if $\beta(s) = 0$, the sequence is rejected.

Example

M is a finite state machine to check that a sequence of symbols consists of letters and numbers in which the initial symbol is a letter:

$$M = (S, \Sigma, S_0, \beta, \delta), \qquad S = \{S_0, S_1, S_2\}, \qquad \Sigma = \{\text{letter, number}\}$$

δ	letter	number		β	
S_0	S_1	S_2		S_0	0
S_1	S_1	S_1		S_1	1
S_2	S_2	S_2		S_2	0

The sequence (letter, number, letter) leads from S_0 to S_1 to S_1 to S_1 and is accepted.

The sequence (number, letter, number) leads from S_0 to S_2 to S_2 and is rejected.   #

Example

$M = (S, \Sigma, s_0, \beta, \delta)$, $\Sigma = \{a, b, c\}$, and $S = \{s_0, s_1, s_2\}$.

δ	a	b	c		β	
s_0	s_0	s_1	s_2		s_0	0
s_1	s_2	s_1	s_1		s_1	1
s_2	s_1	s_0	s_2		s_2	0

The sequence (a, b, c, b) leads from s_0 to s_0 to s_1 to s_1 to s_1 and is accepted.

The sequence (c, b, b, a) leads from s_0 to s_2 to s_0 to s_1 to s_2 and is rejected.   #

Alternatively, δ is a set of next state functions, $\{\delta_\sigma : S \to S \mid \sigma$ in $\Sigma\}$. The effect of a sequence of inputs may be described as follows.

$$\delta(\sigma_1, \sigma_2, \sigma_3) = \delta_{\sigma_1} \circ \delta_{\sigma_2} \circ \delta_{\sigma_3}$$

That is, the effect of the application of the sequence $(\sigma_1, \sigma_2, \sigma_3)$, as a function from $S \to S$, is the composition of the individual functions corresponding to σ_1, σ_2, σ_3. A sequence τ is then accepted just in case $\beta(\delta_\tau(s_0)) = 1$, while τ is rejected just in case $\beta(\delta_\tau(s_0)) = 0$.

Example

For the machine M just described,

$$\delta_a : s_0 \mapsto s_0 \qquad \delta_b : s_0 \mapsto s_1 \qquad \delta_c : s_0 \mapsto s_2$$
$$s_1 \mapsto s_2 \qquad\qquad s_1 \mapsto s_1 \qquad\qquad s_1 \mapsto s_1$$
$$s_2 \mapsto s_1 \qquad\qquad s_2 \mapsto s_0 \qquad\qquad s_2 \mapsto s_2$$

$$\delta_{(a,b)} = \delta_a \circ \delta_b : s_0 \mapsto s_1$$
$$s_1 \mapsto s_0$$
$$s_2 \mapsto s_1$$

$$\delta_{(a,b,c)} = \delta_a \circ \delta_b \circ \delta_c : s_0 \mapsto s_1$$
$$s_1 \mapsto s_2$$
$$s_2 \mapsto s_1$$

$$\delta_{(a,b,c,b)} = \delta_a \circ \delta_b \circ \delta_c \circ \delta_b : s_0 \mapsto s_1$$
$$s_1 \mapsto s_0$$
$$s_2 \mapsto s_1$$

Thus $\delta_{(a,b,c,b)}(s_0) = s_1$ and $\beta(s_1) = 1$, so the sequence (a, b, c, b) is accepted. [Note that the sequence (a, b) and the sequence (a, b, c, b) are equivalent in that both yield the same function $f: S \to S$. This is an important point to which we will return in Chapter 3.]

U. BOOLEAN ALGEBRA

A Boolean algebra consists of a set B, a pair of binary operations θ_1, θ_2, and a unary operation $-$, all defined on B and satisfying the following axioms:

1. θ_1, θ_2 are commutative and associative.
2. (a) There is an element i in B such that, for any $x \in B$,

$$i\theta_1 x = i \qquad i\theta_2 x = B$$

   (b) There is an element 0 in B such that

$$0\theta_1 B = B \qquad 0\theta_2 B = 0$$

3. (a) θ_1 is distributive over θ_2; for any elements x, y, z in B,

$$x\theta_1(y\theta_2 z) = (x\theta_1 y)\theta_2(x\theta_1 z)$$

   (b) θ_2 is distributive over θ_1.

$$x\theta_2(y\theta_1 z) = (x\theta_2 y)\theta_1(x\theta_2 z)$$

4. For each element x in B,

$$(-x)\theta_1 x = i \qquad (-x)\theta_2 x = 0$$

Example

Let $X_1, X_2, X_3, \ldots, X_n$ be subsets of a universal set U. The set of all subsets of $\{X_1, \ldots, X_n\}$, which can be formed from the X_i by union, intersection, and complement, together with the binary operations of $\cup$ and $\cap$, and the unary operation, form a Boolean algebra.

Identifying $\cup$ as θ_1, $\cap$ as θ_2, $\{X_1, \ldots, X_n\}$ as U, ϕ as 0, and complement as $-$, it is easy to check that all the axioms hold.

(a) $\cup$ and $\cap$ are commutative and associative.
(b) (a) $U \cup x = U \qquad U \cap x = x$.
    (b) $\phi \cup x = x \qquad \phi \cap x = \phi$.
(c) (a) $x \cup (y \cap z) = (x \cup y) \cap (x \cup z)$.
    (b) $x \cap (y \cup z) = (x \cap y) \cup (x \cap z)$.
(d) $(\bar{x}) \cup x = U \qquad (\bar{x}) \cap x = \phi$. #

Example

Let $N = p_1 \cdot p_2 \cdot \ldots \cdot p_n$, where each p_i is a prime and, if $i \neq j$, $p_i \neq p_j$. N is the product of distinct primes. Then N, with all its divisors and with the binary operations of greatest common division and least common multiple, and the unary operation taking x to N/x, is a Boolean algebra, with $N = i$ and $1 = 0$; that is, the number 1 is the 0 element of the Boolean algebra.

Again, one can check all the axioms.   #

Each Boolean algebra can be shown as a lattice, with $x_1 \theta_1 x_2$ shown as the lub of x_1 and x_2 and $x_1 \theta_2 x_2$ shown as the glb of x_1 and x_2. U is shown as the maximal element of the lattice, and 0 is shown as the minimal element of the lattice.

Example

Taking three sets A, B, C and all combinations of them under union, intersection, and complement, the following lattice results.

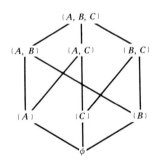

Example

The lattice of the divisors of 30 is:

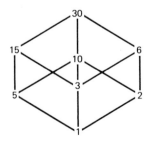

V. SWITCHING ALGEBRA

Definition

   A switching function (or Boolean function) is a function from $\{0, 1\}^h \rightarrow \{0, 1\}$. Alternatively, a switching function is a function from $B^h \rightarrow B$, where B is a set of cardinality 2.

Switching functions are used in the analysis and synthesis of computer switching circuits. Each signal line input to each circuit element is in either a high- or low-voltage state. As a function of the input states, the output is in either a high- or low-voltage state.

ANALYSIS OF A NOR CIRCUIT.

I_1	I_2	$v_1(I_1)$	$v_1(I_2)$	$v_2(I_1)$	$v_2(I_2)$	Output	v_1(Output)	v_2(Output)
High	High	0	0	F	F	Low	1	T
High	Low	0	1	F	T	High	0	F
Low	High	1	0	T	F	High	0	F
Low	Low	1	1	T	T	High	0	F

This physical behavior is interpreted as a symbolic behavior either by assigning arbitrarily the value 0 to a high voltage and the value 1 to a low voltage, or by assigning the logical value T to a low voltage and the value F to a high voltage. The typical name given to the circuit is "NOR" since, based on the assignment of logical values

$$\text{output} = \overline{(I_1 \vee I_2)}$$

That is, the output is the negation of the disjunction of the input.

   Let f_1, f_2 be Boolean functions. $f_1 \oplus f_2$ is defined by $(f_1 \oplus f_2)(x_1, x_2, \ldots, x_n) = f_1(x_1, \ldots, x_n) \oplus f_2(x_1, \ldots, x_n)$, where $\oplus$ is either $+$ or $\cdot$ and where $+$ and $\cdot$ have the following operation tables.

$+$	0	1
0	0	1
1	1	1

$\cdot$	0	1
0	0	0
1	0	1

$-f$ is defined by $-f(x_1, \ldots, x_n) = 0$ if $f(x_1, \ldots, x_n) = 1$, and $-f$ is defined by $-f(x_1, \ldots, x_n) = 1$ if $f(x, \ldots, x_n) = 0$.

Fact.   The set of Boolean functions under the operations of $+$, $\cdot$, and $-$ form a Boolean algebra.

Example

The set of functions from $\{0, 1\}^2$ to $\{0, 1\}$ consists of the 16 functions shown in Table 1-1.

TABLE 1-1 THE SET OF BOOLEAN FUNCTIONS OF TWO VARIABLES.

f	$x_1 = 0$ $x_2 = 0$	$x_1 = 1$ $x_2 = 0$	$x_1 = 0$ $x_2 = 1$	$x_2 = 1$ $x_1 = 1$
f_0	0	0	0	0
f_1	0	0	0	1
f_2	0	0	1	0
f_3	0	0	1	1
f_4	0	1	0	0
f_5	0	1	0	1
f_6	0	1	1	0
f_7	0	1	1	1
f_8	1	0	0	0
f_9	1	0	0	1
f_{10}	1	0	1	0
f_{11}	1	0	1	1
f_{12}	1	1	0	0
f_{13}	1	1	0	1
f_{14}	1	1	1	0
f_{15}	1	1	1	1

Using the rules for combining functions; one can check that

$$f_5 + f_9 = f_{13} \qquad \text{and} \qquad f_5 \cdot f_9 = f_1$$

and that

$$-f_5 = f_{10} \qquad \text{and} \qquad -f_9 = f_6$$

$$(f_5 \cdot f_6 + f_3) \cdot (f_4 + f_8) = f_4 \quad \#$$

Notation.   The Boolean function f_{x_i} defined as

$$f_{x_i}(x_1, \ldots, x_n) = 1 \qquad \text{if} \qquad x_i = 1$$

$$f_{x_i}(x_1, \ldots, x_n) = 0 \qquad \text{if} \qquad x_i = 0$$

is written as x_i. Then, if E_1 denotes f_1 and E_2 denotes f_2, $E_1 + E_2$ denotes $f_1 + f_2$, $E_1 \cdot E_2$ denotes $f_1 \cdot f_2$, and $\bar{E}_1$ denotes $\bar{f}_1$.

Example

$x_1 + x_2$ denotes f_{14} and $x_1 \cdot \bar{x}_2$ denotes f_4.   #

Each expression combining the variables $x_1, \ldots, x_n$ using the operations $+$, $\cdot$, and $-$, denotes a Boolean function. Two expressions E_1 and E_2 are said to be equivalent if they denote the same function. It is often desirable to find the simplest expression, by some criterion, equivalent to a given expression.

Example

$x_1 x_2 \bar{x}_3 + x_1 x_2 x_3 + x_1 \bar{x}_2 x_3$ can be shown to be equivalent to $x_1(x_2 + x_3)$. #

Karnaugh Maps

The simplification of expressions denoting Boolean functions is often important because the Boolean expression generally denotes the *structure* of a logical circuit or program, while the Boolean function describes the *behavior* of the circuit or program. As we have noted, when many different circuits or programs can be used to compute the same function, it is often desirable to select the one that is simplest according to some criterion.

Example

It is desired to find a circuit to compute the function denoted by

$$x_1 \cdot x_2 + x_1 \cdot \bar{x}_2$$

A straightforward circuit realization of this function is

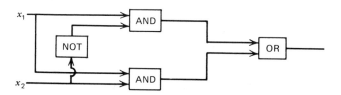

This circuit requires four logical building blocks. However, since $x_1 \cdot x_2 + x_1 \cdot \bar{x}_2$ denotes the same function as x_1, an equivalent, more economical circuit is

$$x_1 \rightarrow$$

The simpler, x_1 circuit requires no logical building blocks.   #

Although the computation of the simplest equivalent expression is, in general, very difficult and no efficient algorithms are known for its solution, there are various methods for computing approximate solutions. Of these, one of the best known is the Karnaugh map method.

A Karnaugh map is a generalization of Venn diagrams to more than three variables. A four-variable Karnaugh map is:

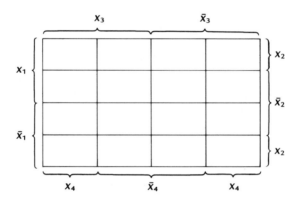

The theoretical basis for the use of Karnaugh maps follows.

Fact.   Every Boolean function of n variables, $x_1, \ldots, x_n$, can be expressed as a sum of products of the variables and their complements.

Proof.   (By induction.)

BASIS ($n = 1$).   In this case, there are only four functions.

$$f_3 = 1 \ \text{if} \ x = 1 \ \text{or} \ x = 0 \ \text{can be denoted} \ x + \bar{x}$$

$$f_2 = 1 \ \text{iff} \ x = 1 \qquad\qquad \text{can be denoted} \ x$$

$$f_1 = 1 \ \text{iff}^5 \ x = 0 \qquad\qquad \text{can be denoted} \ \bar{x}$$

$$f_0 = 0 \qquad \text{is a sum of no terms}$$

---

[5]The letters "iff" mean "if and only if."

Induction. Here we suppose that every function of n variables can be expressed as a sum of products, and we show that, on this assumption, every function of $n+1$ variables can be expressed as a sum of products.

In this case, since x_{n+1} can take only the values 0 and 1, we can form the following functions of n variables from $f(x_{n+1}, x_n, \ldots, x_1)$, $f(1, x_n, \ldots, x_1)$, and $f(0, x_n, \ldots, x_1)$, where $f(1, x_n, \ldots, x_1)$ denotes the function f of n variables when $x_{n+1} = 1$ and $f(0, x_n, \ldots, x_1)$ denotes the function f of n variables when $x_{n+1} = 0$.

Now we can write

$$f(x_{n+1}, x_n, \ldots, x_1) = x_{n+1} \cdot f(1, x_n, \ldots, x_1) + \bar{x}_{n+1} \cdot f(0, x_n, \ldots, x_1)$$

If $x_{n+1} = 1$, then $\bar{x}_{n+1} = 0$, and the second term on the right is $0 \cdot f(0, x_n, \ldots, x_1)$, or 0; if $x_{n+1} = 0$, then $\bar{x}_{n+1} = 1$ and the first term on the right is $0 \cdot f(1, x_n, \ldots, x_1)$, or 0.

Now, by distributing the x_{n+1} and $\bar{x}_{n+1}$ factors, we obtain an expression as a sum of products.

Example

To express $f(x_2, x_1) = x_1 + x_2$ as a sum of products,

$$f(x_2, x_1) = x_2 \cdot f(1, x_1) + \bar{x}_2 \cdot f(0, x_1)$$

where	$f(1, x_1) = x_1 + 1$	substituting 1 for x_2 in $f(x_2, x_1) = x_2 + x_1$, simplifying $x_1 + 1$
	$= 1$	in a Boolean algebra
and where	$f(0, x_1) = x_1 + 0$	substituting 0 for x_2 in $f(x_2, x_1) = x_2 + x_1$, simplifying $x_1 = 0$.
	$= x_1$	

Now $f(1, x_1) = x_1 \cdot f(1, 1) + \bar{x}_1 \cdot f(1, 0)$.

where	$f(1, 1) = 1$	substituting 1 for x_1 in $f(1, x_1) = 1$
and where	$f(1, 0) = 1$	substituting 0 for x_1 in $f(1, x_1) = 1$
Thus,	$f(1, x_1) = x_1 + \bar{x}_1$.	

Furthermore $f(0, x_1) = x_1 \cdot f(0, 1) + \bar{x}_1 \cdot f(0, 1) + \bar{x}_1 \cdot f(0, 0)$,   and $f(0, 1) = 1$, substituting 1 for x_1 in $f(0, x_1) = x_1$ and $f(0, 0) = 0$, substituting 0 for x_1 in $f(0, x_1) = x_1$, thus $f(0, x_1) = x_1$.

Substituting in $f(x_2, x_1) = x_2 \cdot f(1, x_1) + \bar{x}_2 \cdot f(0, x_1)$ yields

$$f(x_2, x_1) = x_2(x_1 + \bar{x}_1) + \bar{x}_2 \cdot (x_1)$$

$$= x_2 x_1 + x_2 \bar{x}_1 + \bar{x}_2 x_1 \qquad \#$$

Each square in the Karnaugh map represents a product of the variables and their complements; each Boolean function can be represented by the set of squares whose products appear in the sum-of-products form.

Example

$x_1 x_2 x_3 x_4 + x_1 \bar{x}_2 x_3 x_4 + x_1 x_2 \bar{x}_3 x_4$ is represented by the following Karnaugh map.

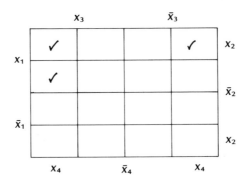

We have checked the squares corresponding to products in the sum-of-products form of the expression.   #

The use of the Karnaugh map for simplification of expressions is based on the recognition that certain (not necessarily contiguous) areas of the Karnaugh map are expressed more compactly in this way than as a sum of products.

Example

Continuing the previous example and noting that

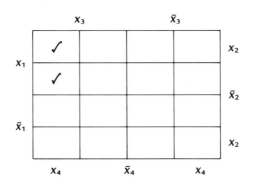

can be denoted $x_1x_3x_4$ and

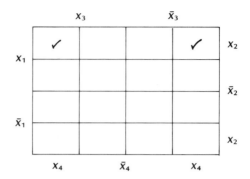

can be denoted $x_1x_2x_4$, we obtain a simplified expression, $x_1x_2x_4 + x_1x_3x_4$, that denotes the same function.   #

Exercise.   The dual of the sum-of-products form is the product-of-sums form. [$f(x_1, x_2) = x_2x_2$ can be expressed in product-of-sums form as $f(x_1, x_2) = (x_1 + x_2)(x_1 + \bar{x}_2)(\bar{x}_1 + x_2)$.] Show that any Boolean function can be denoted as a product of sums, and show how the Karnaugh map can be used for the product-of-sums form.

Decision Tables

Another application for the simplification of Boolean expressions is in the implementation of decision tables. A decision table is a program specification consisting of a set of logical conditions and a list of actions to be taken when the appropriate combination of the conditions is true. These are usually represented in a table, where each column indicates a set of conditions and the corresponding actions.

Example

Conditions				
C_1	T	F	T	T
C_2	T	—	F	T
C_3	F	—	—	T

Actions				
a_1	✓	✓		✓
a_2		✓		✓

In this decision table, there are three conditions, C_1, C_2, and C_3, and two actions, a_1 and a_2. In the first column we see that if C_1 and C_2 are true and C_3 is false, then action a_1 is to be taken. In the second column, we see that if C_1 is false, actions a_1 and a_2 are both to be taken.   #

The methods of analysis of Boolean functions can be used with decision tables to insure that every set of possible conditions has an associated action, so that the condition specification is *complete*, and that two different sets of actions are not specified for the same conditions, so that the specification is *consistent*.

Example

Continuing our previous example, we may use a three-variable Karnaugh map to verify the consistency and completeness of the conditions statement.

	C_1			
C_3	column 3	column 4	column 2	column 2
$\bar{C_3}$	column 3	column 1	column 2	column 2

$$\bar{C_2}$$

In each square of the Karnaugh map, we can indicate which columns cover that square. If every square is covered by exactly one column, we have completeness and consistency.   #

REVIEW

The key concepts introduced in this chapter are sets, relations, and functions. These abstractions and the operations that combine and relate them are developed.

An example combining some of these concepts is the finite state machine. A finite state machine is itself an abstraction that can be used to represent a number of different concrete situations.

PROBLEMS

E 1. Write concise descriptions of:
(a) The set of nonnegative numbers.
(b) The set of positive numbers divisible by 7.
(c) The set of integers not greater than 100 that are not perfect squares of integers.

P 2. A sequence S is a shuffle of sequences A and B if the elements of A and B occur in S; each element of S is either an element of A or an element of B; and if p occurs before q in A or B, then p occurs before q in S.

Example
$(1, 2, a, b, 3)$ is a shuffle of (a, b) and $(1, 2, 3)$.   #

Assuming that all elements are distinct and that A has 3 elements and B has 4 elements, how many distinct shuffles of A and B are there?

E 3. Write concise descriptions, using set notation, of:
(a) The sets of integers that are perfect squares.
(b) The sets of positive integers that yield a remainder of 5 when divided by 11.
(c) The numbers that satisfy both conditions, that is, are perfect squares and are positive integers yielding a remainder of 5 when divided by 11.

E 4. Let $S_1 = \{1\}$, $S_2 = \{1, 2, 3\}$, $S_3 = \{2, 3\}$, $S_4 = \{1, 2, 3, 2\}$, and $S_5 = \{2, 3, 1\}$.
(a) How many members does each of S_1 to S_5 have?
(b) Which of these sets is the same as S_2?
(c) Which of these sets is a subset of S_2? A proper subset?

P 5. Propose two alternative definitions of *subsequence* that are not equivalent and yet each provides $A = B$ if and only if A is a subsequence of B and B is a subsequence of A.

E   6.   Write a computer program to check if an explicitly specified set A is a subset of an explicitly specified set B. How
do you represent sets in your program?

E   7.   Calculate $P(\{a, b\})$ and $P(P(\{a, b\}))$.

E   8. $\emptyset$ is not a member of every set. Suppose S is a set, and $\emptyset$
is not in S. Is $\emptyset$ in $P(S)$? Justify your answer.

P   9.   (a) Write a recursive program to compute $P^k(\emptyset)$, and
print out $P^6(\emptyset)$.
(b) Estimate the number of symbols needed to list the
members of $P^k(\emptyset)$.

E 10.   Let $A = \{1, 2, 3, 4, 5, 6\}$, $B = \{2, 4, 6, 8\}$,   and $C = \{x \mid x = n^2 + 1, \, n \geq 0\}$.
(a) Find $A \cap B$, $A \cap C$, $B \cap C$, and $A \cap B \cap C$.
(b) Find $A \cup B$, $A \cup C$, and $B \cup C$.
(c) Find $A - B$, $B - A$, $A \oplus B$, and $A \oplus C$.

E 11.   Let $A = \{1, 2, 3, 4\}$, $B = \{2, 4, 6\}$,   and $C = \{1, 3, 5\}$.   Calculate:
(a) $A \cup B$.
(b) $A \cup C$.
(c) $B \cap C$.
(d) $B - A$.
(e) $C - A$.

P 12.   Let $A = \{1, 3, 5\}$, $B = \{2, 4, 6\}$, and $C = \{1, 4, 5\}$, and the
universal set is $\{1, 2, 3, 4, 5, 6\}$. Compute all the different
sets that can be formed from A, B, and C using union,
intersection, and complement.

E 13.   Prove that symmetric difference is commutative.

E 14.   Prove that symmetric difference is associative.

P 15.   (a) Give two different possible interpretations of $A \cap B \cup C$.
(b) Illustrate your different interpretations for $A = \{1, 2, 5, 7\}$, $B = \{2, 4, 6\}$, and $C = \{3, 4, 5\}$.

E 16. Show the Hasse diagram for the set of sets computed in Problem 12.

E 17. (a) Show the set representations of the sequences $(\{1\}, 2)$, $(1, \{2\})$, and $(\{1\}, \{2\})$.
(b) Why isn't the set $\{a, \{b\}\}$ an adequate representation of (a, b)? (Show an example where, under this representation, two different ordered pairs would yield the same set.)

E 18. Show the set form of the following sequences.
       (a) $(1, 2)$.
       (b) $(1, \{1\}, \{\{1\}\})$.
       (c) $(1, 2, 3)$.

E 19. Consider the sequences $(1, 2)$ and $(2, 1)$.
       (a) Show their set representations.
       (b) Does the set union of their representations represent a sequence?
       (c) What about their set intersection?

E 20. Let $A = \{1, 2, 3, 4\}$ and $B = \{4, 5, 6, 7, 8\}$. Show the relation $xRy \subseteq A \times B$ where:
       (a) xRy iff x divides y.
       (b) xRy iff gcd $(x, y) = 1$.
       (c) xRy iff exactly one of x and y is a prime.

E 21. Let R_1, $R_2 \subseteq U = \{1, 2, 3, 4\} \times \{1, 2, 3, 4\}$, $R_1 = \{(1, 2), (2, 3), (3, 4)\}$, and $R_2 = \{(1, 3), (2, 4)\}$. List the different relations that can be formed from U, R_1, and R_2 using set operations.

E 22. All the following relations are considered to be subsets of $\{1, 2, 3, 4\} \times \{1, 2, 3, 4\}$. For each row and column, if the relation corresponding to the column label satisfies the property indicated by the row label, insert a check. (*Example*. In the first row, the leftmost space should be checked if $\emptyset$ is reflexive.) (*Note*. E is the *identity* relation. $U = \{1, 2, 3, 4\} \times \{1, 2, 3, 4\}$ is the universal relation.)

	$\emptyset$	E	U	$\{(2, 3)\}$
Reflexive				
Irreflexive				
Symmetric				
Antisymmetric				
Asymmetric				
Transitive				
Equivalence relation				
Partial order				

E 23. Classify the following relations (i.e., state whether they are reflexive, irreflexive, symmetric, antisymmetric, asymmetric, transitive, equivalence relations, partial orders, linear orders, well-orderings). List all that apply and no others. Assume that the universe is $\{1, 2, 3, 4\} \times \{1, 2, 3, 4\}$.

(a) $\{(1, 1), (1, 3), (3, 3), (2, 2), (4, 4)\}$.
(b) $\{(1, 1), (2, 2), (3, 3), (4, 4)\}$.
(c) $\{(1, 1), (1, 3), (3, 1), (3, 3)\}$.
(d) $\{(1, 2), (2, 3), (3, 4), (1, 3), (2, 4), (3, 3)\}$.
(e) $\{(1, 2), (2, 1), (3, 4), (4, 3)\}$.

E 24. Let $A = \{1, 2, 3\} \times \{1, 2, 3, 4\}$. We have the following equivalence relation, E, on A.

$$(x, y)E(u, v) \qquad \text{iff} \qquad x + y = u + v.$$

Compute the partition associated with E.

E 25. A is the same as in Problem 24, but now

$$(x, y)E'(u, v) \qquad \text{iff} \qquad |x - y| = |u - v|$$

Compute the partition associated with E'.

P 26. (a) Describe an algorithm to compute the number of partitions of a finite set.
(b) Compute the number of partitions of sets containing 3, 4, and 5 elements.

E 27. Compute all partitions of the set $\{a, b, c, d\}$.

E 28. For the partition, $\pi = \{1, 2; 3, 4; 5\}$, compute the corresponding equivalence relation.

P 29. Show that aRb if and only if a and b have the same remainder when divided by 3 is an equivalence relation. What is the corresponding partition?

E 30. Let A and B be finite sets. If A has $|A|$ members and B has $|B|$ members, how many members does $A \times B$ have? How many members does $P(A \times B)$ have?

P 31. Show that if A, B are enumerable, $A \times B$ is enumerable.

P 32. Show that "is a subset of" is a partial order on the power set of set S.

* 33. (a) Let $P = [S, \leq]$ and $Q = [T, \leq]$ be two posets. Define $[S \times T, \leq] = P \times Q$ by having $(s, t) \leq (s', t')$ mean that $s \leq s'$ in P and $t \leq t'$ in Q. Show that $P \times Q$ is a poset.
(b) Let P and Q be as in part a. Define $P \times Q = [S \times T, \leq]$ and $(s, t) = (s', t')$ if either $s \leq s'$ or $s = s'$ and $t \leq t'$. Prove that $P \times Q$ is a poset.

P 34. Given any finite, partially ordered set, S, with ordering relation, R, one can always construct a linear ordering relation, L, so that if xRy, then xLy. (This is sometimes referred to as topological sorting.)

Example

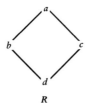

R

L

Note that L is not unique. Also, L may be represented as a sequence—$(a\,b\,c\,d)$.   #

For the partial order

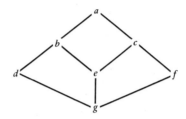

there are 16 possible choices of L. Give *three* different linear orderings (showing them as sequences), including at least one where b is not the second or third element in the sequence.

P 35. Explain why every finite partial order can be topologically sorted (see Problem 34). What is the difficulty with in-finite partial orders?

P 36. Let R be a reflexive relation that is not necessarily sym-metric or transitive.
(a) Show that $R \circ \check{R}$ is reflexive.
(b) Show that $R \circ \check{R}$ is symmetric.
(c) Is $R \circ \check{R}$ transitive? Prove or give a counterexample.
(d) What can you conclude about $R \circ \check{R}$? Is it an equivalence relation?
(e) What about the transitive closure of $R \circ \check{R}$ where R is reflexive? Is it an equivalence relation?

P 37. (a) Show that if $A \subseteq B$, then $\check{B} \supseteq \check{A}$.
(b) We know that $A \circ B = \check{B} \circ \check{A}$. Using this fact, show that the converse of a transitive relation is transitive.
(c) Using the preceding, show that the converse of an equivalence relation is an equivalence relation.

E 38. Which of the following properties of relations are preserved under complement? Under converse?
(a) Reflexivity.
(b) Symmetry.

(c) Asymmetry.

(d) Antisymmetry.

(e) Transitivity.

E 39. Determine whether each of the following relations defined by ordered pairs (m, n) of natural numbers has each of the properties given in Problem 38.

 (a) m is divisible by n.

 (b) $m + n$ is odd.

 (c) $m + n$ is even.

 (d) $m \circ n$ is even.

 (e) $m \circ n$ is odd.

 (f) $m \circ n \leqq m^2 \circ m^2$.

P 40. Show that the converse of a partial order is a partial order.

P 41. Show that the composition of symmetric relations is not necessarily symmetric. (*Hint.* Find a counterexample.)

E 42. Compute the composition of

$$R_1 = \{(1, 1), (1, 3), (2, 2), (2, 3), (3, 1), (3, 4)\}$$

and

$$R_2 = \{(1, 3), (2, 1), (2, 3), (4, 1), (4, 2)\}$$

E 43. (a) For the R_1 and R_2 in Problem 42, compute $R_2 \circ R_1$.

 (b) Compute $\check{R}_2 \circ \check{R}_1$.

P 44. Prove that a permutation that is transitive is an equivalence relation.

P 45. (a) Show that the composition of any two injections $f: S \rightarrow T$ and $g: T \rightarrow U$ is an injection.

 (b) Do the same for surjections.

* 46. Write a program to compute the converse of a permutation. (See if you can do this using only a fixed finite amount of storage in addition to the storage of the permutation itself, independent of the size of the permutation.)

P 47. Show that the composition of two permutations is a permutation.

P 48. Show that the converse of a permutation, p, is a permutation, p'. Furthermore, show that $p' \circ p = p \circ p' = E$, where E is the identity permutation.

P 49. The cycle length of a permutation, p, is the smallest positive integer, k, such that $p^k = E$, where E is the identity permutation. Write a program to compute the cycle length of a permutation.

P 50. Let X, Y, and Z be subsets of U with $Y \cap Z = \emptyset$ and $Y \cup Z = X$. Construct a bijection, b.

$$b: P(X) \leftrightarrow P(Y) \times P(Z)$$

E 51. Give examples of relations having the following properties; in each case state the range and domain.
    (a) Symmetric, reflexive, nontransitive.
    (b) A function that is $1-1$ but not onto.
    (c) A function that is onto but not $1-1$.
    (d) *A partial order that is an equivalence relation.*
    (e) Antisymmetric but not asymmetric.

E 52. $E = \begin{pmatrix} 1 & 2 & 3 & 4 \\ 5 & 6 & 7 & 8 \end{pmatrix}$ and $F = \begin{pmatrix} 1 & 3 \\ 5 & 7 \\ 9 & 11 \\ 13 & 15 \end{pmatrix}$. Compute $E_+^\times F$.

E 53. Let $E = \begin{pmatrix} 1 & 3 & 5 & 7 \\ 2 & 5 & 8 & 11 \end{pmatrix}$ and $F = \begin{pmatrix} 4 & 2 \\ 3 & 6 \\ 2 & 1 \\ 1 & 5 \end{pmatrix}$.

    (a) Compute $E \, _+^+ \, F$.

    (b) Compute $F \, _+^\times \, E$.

E 54. Consider the generalized matrix product $A_{\theta_2}^{\theta_1} B$. Find an example where $(A_{\theta_2}^{\theta_1} B)_{\theta_2}^{\theta_1} C \neq A_{\theta_2}^{\theta_1} (B_{\theta_2}^{\theta_1} C)$.

P 55.   In Problem 52, what condition is necessary for the generalized product to be associative?

P 56.   *Definition. A relation R is said to be circular if aRb and bRc imply cRa.*
      (a) Give a matrix test for circularity.
      (b) Show that R is an equivalence relation if and only if R is reflexive, symmetric, and circular.

P 57.   (a) How many functions from a set of n elements to a set of m elements are surjective, $n \geqq m$?
      (b) How many functions from a set of n elements to a set of m elements are injective, $n \leqq m$?

E 58.   Let $A \subseteq \{1, 2, 3, 4\} \times \{1, 2, 3, 4\}$, $A = \{(1, 1), \ (2, 3), \ (3, 2),$ $(3, 4), \ (4, 1)\}$, $B \subseteq \{1, 2, 3, 4\} \times \{1, 2\}$, and $B = \{(1, 1), \ (3, 1),$ $(3, 2)\}$.
      (a) Show the matrix forms of A and B, and compute the matrix form of $A \cdot B$.
      (b) Show the matrix forms of A^T and B^T, the transposes of A and B, respectively, and compute the matrix form of $B^T \cdot A^T$.

E 59.   For the relation, A, of Problem 58, compute the reflexive closure, the symmetric closure, and the transitive closure.

E 60.   Let π_1, π_2 be the following permutations on $\{1, 2, 3, 4, 5\}$.

$$\pi_1: 1 \mapsto 2 \qquad \pi_2: 1 \mapsto 3$$
$$2 \mapsto 3 \qquad 2 \mapsto 5$$
$$3 \mapsto 1 \qquad 3 \mapsto 1$$
$$4 \mapsto 5 \qquad 4 \mapsto 2$$
$$5 \mapsto 4 \qquad 5 \mapsto 4$$

      (a) Show the matrices of π_1 and π_2.
      (b) What is π_1^{-1}?
      (c) Compute $\pi_1^{-1} \circ \pi_2$.

E 61. Let **M** be the matrix of an equivalence relation. Write a program to rearrange **M** in standard form.

E 62. Compute the transitive closure of A, where

$$A = \{(1, 2), (2, 3), (3, 4), (4, 2)\}$$

P 63. Write a program to check if a relation R is transitive.

P 64. The product of two lattices $[L_1, \wedge_1, \vee_1]$ and $[L_2, \wedge_2, \vee_2]$ is defined to be $[L_1 \times L_2, \wedge_3, \vee_3]$ where $(a, b) \wedge_3 (c, d) = (a \wedge_1 c, b \wedge_2 d)$ and $(a, b) \vee_3 (c, d) = (a \vee_1 c, b \vee_2 d)$. Show the product of the two lattices:

* 65. Show that the power set is a lattice under union and intersection, and draw the diagram of the lattice of $P(\{a, b, c, d\})$.

E 66. Show the lattice of $[S, \gcd, \mathrm{lcm}]$ where $S = \{x \mid x$ is a divisor of 210$\}$.

P 67. (a) How many relations are there from a set A with $|A|$ elements to a set B with $|B|$ elements?
(b) How many functions are there from $P(A)$ to $P(B)$?

P 68. A (finite) *Boolean algebra* is a (finite) lattice satisfying some additional conditions. Every finite lattice has a maximal element I, such that $I \vee x = x \vee I = I$ for each element x in the lattice, and a minimal element 0, such that $0 \wedge x = x \wedge 0 = 0$ for each element x in the lattice. (*Note.* $x \vee y$ denotes the lub of x and y.) For each element x, there is a unique element $\bar{x}$, such that $x \wedge \bar{x} = 0$ and $x \vee \bar{x} = I$. The lattice of a Boolean algebra is distributive: that is, it satisfies the equations:

$$x \wedge (y \vee z) = (x \wedge y) \vee (x \wedge z)$$
$$(x \vee y) \wedge z = (x \wedge z) \vee (y \wedge z)$$
$$x \vee (y \wedge z) = (x \vee y) \wedge (x \vee z)$$
$$(x \wedge y) \vee z = (x \vee z) \wedge (y \vee z)$$

(a) Prove that every finite lattice has a maximal element and a minimal element.

(b) Show that the set of all subsets of a finite set constitutes a Boolean algebra, where the partial order underlying the lattice is set containment.

(c) Draw a lattice diagram for the lattice of all subsets of $\{1, 2, 3, 4\}$.

(d) Explain why the following lattice is not a Boolean algebra.

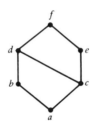

P 69. Show the lattice of subsets of $\{a, b, c\}$ under the ordering $\subseteq$ and give a topological sorting of this lattice.

E 70. Let $R \subseteq \{1, 2, 3, 4\} \times \{a, b, c, d\}$ and $R = \{(1, a), (1, c), (2, a), (2, d), (3, a), (3, b), (3, c)\}$. Show the function associated with R as a mapping from $P(\{1, 2, 3, 4\})$ to $P(\{a, b, c, d\})$.

E 71. Describe a finite state machine that accepts a sequence if and only if it is the binary representation of a number divisible by 3 [e.g., $(11001)_2 = (25)_{10}$ is not divisible by 3 and is not accepted; while $(10101)_2 = (21)_{10}$ is divisible by 3 and is accepted.]

* 72. Given a finite state machine, M, let s_i, s_j be a pair of states. We have $s_i R s_j$ if for every sequence, σ, the sequence σ leads from s_i to a final state if and only if σ

leads from s_i to a final state. Prove that R is an equivalence relation.

E 73. Let $M = (S, \Sigma, s_0, \beta, \delta)$, where $\Sigma = \{a, b, c\}$ and $S = \{s_0, s_1, s_2\}$, with

δ	a	b	c
s_0	s_0	s_1	s_2
s_1	s_2	s_2	s_1
s_2	s_1	s_0	s_1

(a) Compute $\delta_{abbaccab}$.

(b) Is there a shorter sequence than *abbaccab* that has the same state mapping?

P 74. Show the lattice of the set of divisor of 12 under the operations of greatest common divisor and least common multiple. Does this set of divisors and operations constitute a Boolean algebra? Explain.

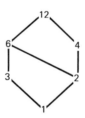

E 75. Find a Boolean expression equivalent to $\bar{x}_1 x_2 x_3 x_4 + \bar{x}_1 \bar{x}_2 x_3 x_4 + \bar{x}_1 x_2 x_3 \bar{x}_4$ that is simpler than this expression. State your criterion for simplicity.

SUGGESTED PROGRAMMING EXERCISES

Problems 6, 46, 49, and 63 are programming problems. So are the following exercises.

1. Write programs to perform the following set operations:
    (a) Union.
    (b) Intersection.

(c) Complement.

(d) Power set.

2. Write and run the sieve of Eratosthenes program to find all primes less than 1000.

3. Write programs to check the following properties of relations.
   (a) Reflexivity.
   (b) Symmetry.
   (c) Antisymmetry.
   (d) Asymmetry.
   (e) Transitivity.

4. (a) Write a program to perform the generalized matrix product.
   (b) Specializing to the arithmetic operations + and ×, show that your program computes the usual matrix multiplication.
   (c) Specializing to the logical operations ∨ and ∧, show that your program computes the composition of relations.

2

directed graphs

SUMMARY

Directed and labeled graphs are used to represent models of programs and problem statements; a familiar example is the program flowchart. It is clear that the primary value of such a graphic representation is that it facilitates our (human) understanding, even though we do not clearly understand why our minds seem to be more efficient in processing such graphic forms. Yet once we have decided that such graphs are desirable we are confronted with a dual problem.

1.   Given a graph representing a problem situation or a relationship among facts, how do we summarize the information contained in the graph for processing by the computer? How do we state algorithms for processing graphs in computer form?

2.   Given a set of facts and relationships among these facts that is present in nongraphic form—perhaps as a set of functions and relations—how do we arrange this information in the form of a graph for human appreciation?

In this chapter, we are concerned with both the representation of graphs in a form amenable to the computer and the representation of information in the form of a graph.

Once you appreciate that every problem in computer science can be phrased as a problem in graphs, you can understand that it is not feasible to cover graph theory exhaustively, even in the limited computer science content. This chapter, then, answers the preceding two questions and provides sufficient examples to motivate the student to further study of graphs.

A. DIRECTED GRAPHS

Definition

A (relational) directed graph (digraph, *for short*) is (described as) a set of vertices, V, and a set of edges, $E \subseteq V \times V$. If a, b are vertices and (a, b) is in E, there is said to be an edge from a to b.

Figure 2-1

(a)

$$V = \{a, b, c\}$$
$$E = \{(a, a), (a, b), (b, c), (c, b)\}$$

(b)

Example
Figure 2-1 illustrates the definition. In Figure 2-1a a directed graph *G* is shown; in Figure 2-1b its description as a set of vertices and edges is shown.   #
Observe that in the definition there can be at most one edge from vertex *a* to vertex *b*, since the edges are describable as a set of ordered pairs of vertices. (There may be an edge from *a* to *b* and an edge from *b* to *a*.) Graphs without this restriction are called *multigraphs*.

Definition
If $(a, b) = e$ is an edge in *G*, *a* is said to be the source of *e* and *b* is said to be the target of *e*.

Fact. Every (relational) directed graph can be represented as a relation and every relation can be represented as a directed graph.
Relational facts and their characterizations have visual representations as graphs. The construction each way is straightforward.

Definition
A pair of vertices *a*, *b* is said to be connected in *G* if (a) (a, b) is in *E*, (b) (b, a) is in *E*, or (c) *there is a vertex c, and a is connected to c and b is connected to c.*

Note.
(a) If *R*, *G* are the relation and graph, respectively, that correspond to each other, then *a* may be connected to *b* in *G*

but not in R; on the other hand, if a is connected to b in R, it will also be connected in G. Thus the term "connected" has a meaning that depends on whether it is applied to graphs or relations, and it denotes different concepts.

(b) The graph definition of connectedness is a recursive definition.

(c) Connectedness in a directed graph is symmetric and transitive but not necessarily reflexive. (An isolated vertex is *not* connected to itself.)

Definition

A graph *is said to be connected if every pair of vertices is connected.*

Definition

Let G *be a directed graph* (E, V). *Then* G', *the* undirected graph *corresponding to* G, *may be described as* $G' = (E', V')$, *where* $V' = V$; *that is, the set of vertices is the same, and there is an edge between a, b in* G' *if* (a, b) *is in* G *or if* (b, a) *is in* G.

Example

Figure 2-2a shows a directed graph, and Figure 2-2b shows the corresponding undirected graph.   #

Note that there is a unique undirected graph for each directed graph but, given an undirected graph, it corresponds to possibly many directed graphs.

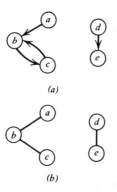

(a)

(b)

Figure 2-2

Both the notions of connectedness for vertices and for graphs are seen to be more naturally associated with the un-directed graph.

Examples

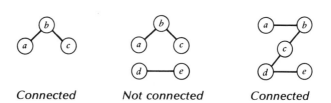

Connected   Not connected   Connected   #

Definition
Let G be a directed graph. If a is any vertex of G, then λ is a (null) path *from a to a*. If (a, b) is an edge of G, there is a path (a, b) *from a to b*. Furthermore, if π_1 is a path from a to c and π_2 is a path from c to b, there is a path $\pi_1 \cdot \pi_2$ from a to b. If π is a path from a to b, a is called the *source* of π and b is the *target* of π.

Note that if π is a path from a to a, $\pi \cdot \pi$ is also a path from a to a. Thus, letting

$$\pi^1 = \pi$$

and

$$\pi^{n+1} = \pi \cdot \pi^n \qquad n > 1$$

we have that $\{\pi^n \mid n \geq 1\}$ is a set of paths from a to a.

Example
In Figure 2-3 there are many paths from vertex a to any other vertex. Thus $(a, b)((b, d)(d, c)(c, b))^n$ is a path from a to b for each $n \geq 0$. Similarly, there is an infinite set of paths from d to g.   #

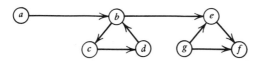

Figure 2-3

The preceding definition of path is recursive. An equivalent nonrecursive definition is that a path is either the null path from vertex to itself or, as a sequence of one or more edges $E_1, E_2, \ldots, E_n$, where the source of E_{i+1} is the same as the target of E_i, $1 \leq i \leq n-1$, in which case it is a path from the source of E_1 to the target of E_n. (Informally, the length of a path is the distance in edges, each edge counting as 1.) The length of the null path is 0.

Definition

A cycle *in a directed graph is a path of length* ≥ 1 *from a vertex to itself.*

If there is a cycle in a graph G from a vertex x to itself, there is an infinite set of paths from x to itself. Conversely, if there is an infinite set of paths from a vertex x to itself, there is at least one cycle from x to itself.

Fact.   Let G be a graph with n vertices. If there is a path from a to b in G, there is a path whose length is less than n.

Proof.   If l is the path length of π, π passes through $l-1$ vertices and, together with its source and target, π touches $l+1$ vertices.

We claim that if the path length is n or more, we can find a shorter path.

Let π be a path from a to b. If the path length is n or greater, π must have touched some vertex, c, twice. Thus π is of the form $\pi_1 \pi_2 \pi_3$, where π_1 is a path from a to c, π_3 is a path from c to b, and π_2 is a path from c to c. But then $\pi' = \pi_1 \pi_3$ is also a path from a to b.   #

Note that in the proof we have implicitly allowed π_1 or π_3 to be a "null path" (i.e., a path of length 0).

Definition

G is strongly connected *if, for every pair of vertices a and b, there is a path from a to b.*

Example

G_1 (shown in Figure 2-4 without the dotted line) is connected but not strongly connected. G_1' (Figure 2-4 with the dotted line) is strongly connected.   #

Figure 2-4

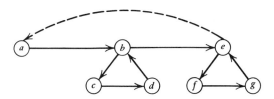

B. GRAPHS AND MATRICES

Given any relational directed graph, G, we may represent it in matrix form by the same matrix that is used to represent its relation. The relational matrix of a graph is also called the *adjacency* matrix.

Example

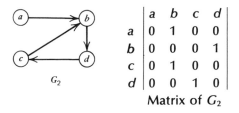

	a	b	c	d
a	0	1	0	0
b	0	0	0	1
c	0	1	0	0
d	0	0	1	0

Matrix of G_2 #

Note. We can tell that G_2 is the graph of a function because there is one and only one edge leading away from each vertex. This corresponds to the matrix having only a single "1" in each row.

Notation. M_G denotes the matrix of G.

Observe that M_{G_2} contains information on all paths of length 1 (i.e., edges) in G_2 and $M_{a,b} = 1$ if and only if there is a path of length 1 from a to b. $[M_{G_2}]^2 = M_{G_2} \overset{v}{\underset{\wedge}{}} M_{G_2}$ is the matrix of $R_{G_2} \circ R_{G_2}$ where R_{G_2} is the relation of G_2, and $\circ$ denotes composition. $[M_{G_2}]^2$ contains information on all paths of length 2 in G_2.

Definition

$$[M_G]^k = [M_G]^{k-1} \overset{v}{\underset{\wedge}{}} M_G \qquad k \geq 2.$$

$[M_G]^k$ *is the matrix of the relation of G composed with itself $k - 1$ times; it contains information on all paths of length k in G.*

We define $[M_G]^+ = \bigcup\limits_{i=1}^{\infty} [M_G]^i$. $[M_G]^+$ is the matrix of the transitive closure of the relation corresponding to G and contains information on all paths of length ≥ 1. $[M_G]_{i,j}{}^+ = 1$ if and only if there is some path from vertex i to vertex j.

Here the graphic interpretation of a relation is quite helpful. In G there is an edge from a to b just in case aRb in the relation R corresponding to G. There is then a path of length k in G from a to b just in case a is related to b in R^k, the k-fold composition of R with itself. This can be seen by an inductive argument:

BASIS. $M_G = M_G{}^1$ contains information on all edges and, therefore, on all paths of length 1.

Induction.   Assume that $M_G{}^k$ contains information on all paths of length k. Then $(M_G{}^{k+1})_{a,b}$ is 1 if and only if, for some c, $(M_G{}^k)_{a,c}$ is 1 and $(M_G)_{c,b}$ is 1. But $(M_G{}^k)_{a,c} = 1$ if and only if there is a path of length k from a to c, and $(M_G)_{c,b} = 1$ if and only if there is a path of length 1 (i.e., an edge) from c to b. But if there is a path of length k from a to c, and a path of length 1 from c to b, then there is a path of length $k + 1$ from a to b.   #

Now the graph interpretation of transitivity is that a relation is transitive if, in the graph of the relation, whenever there is a path from vertex a to vertex b, there is an edge (a, b). The transitive closure of G has an edge wherever there is a path of length 1, or greater, in G.

Note.   Let n be the number of vertices of a graph. If there is a path between vertices i and j there will be a path of length $\leq n$. Thus, we may compute the transitive closure, using only finitely many powers of M_G, by

$$[M_G]^+ = \bigcup\limits_{i=1}^{\infty} [M_G]^i = \bigcup\limits_{i=1}^{n} [M_G]^i$$

where M_G is an $n \times n$ matrix.

Example

We continue with M_{G_2} from the previous example.

$$[M_{G_2}]^2 = M_{G_2} \overset{\vee}{\underset{\wedge}{}} M_{G_2} = \begin{bmatrix} 0 & 0 & 0 & 1 \\ 0 & 0 & 1 & 0 \\ 0 & 0 & 0 & 1 \\ 0 & 1 & 0 & 0 \end{bmatrix}$$

$M_{G_2}^2$ contains information on all paths of length 2.

$$[M_{G_2}^2]^3 = [M_{G_2}]^2 \overset{\vee}{\wedge} M_{G_2} = \begin{bmatrix} 0 & 0 & 1 & 0 \\ 0 & 1 & 0 & 0 \\ 0 & 0 & 1 & 0 \\ 0 & 0 & 0 & 1 \end{bmatrix}$$

$[M_{G_3}]^3$ contains information on all paths of length 3.

The union of these matrices gives us *all* the path information for paths of any length.

$$[M_{G_2}]^+ = M_{G_2} \cup [M_{G_2}]^2 \cup [M_{G_2}]^3 = \begin{bmatrix} 0 & 1 & 1 & 1 \\ 0 & 1 & 1 & 1 \\ 0 & 1 & 1 & 1 \\ 0 & 1 & 1 & 1 \end{bmatrix} \quad \#$$

C. WARSHALL'S ALGORITHM

Warshall's algorithm is an algorithm for computing the connectivity matrix, **C**, which is generally much more efficient than taking successive powers of the adjacency matrix, **A**. We develop this algorithm next.

Definition

If π is a path, a sequence of zero or more edges in G, $\pi = E_1 E_2 \ldots E_k$, then the set of vertices interior to π is $\{x | x$ is the source of $E_i, i > 1\}$.

Example

In the path $(1, 2)(2, 1)(1, 3)$, $\{1, 2\}$ is the set of interior vertices, since $\pi = E_1 E_2 E_3$ with $E_1 = (1, 2)$, $E_2 = (2, 1)$, and $E_3 = (1, 3)$. The set of interior vertices is defined as

$$\{x | x \text{ is the source of } E_2 \text{ or } E_3\} \quad \#$$

We assign to each vertex in G the number of the row and column of **A** associated with that vertex. The computation of **C**, the connectivity matrix, is done iteratively based on intermediate matrices, $C^{(k)}$, where the matrix $C^{(k)}$ is the matrix of vertices connected by paths having only vertices numbered $\leq k$ in their interior. The computation of $C_{ij}^{(k)}$ is iterative with $\boldsymbol{C}^{(0)} = \boldsymbol{A} \vee \boldsymbol{I}$. $C_{ij}^{(0)}$ is 1 if there is an edge from i to j or if $i = j$; therefore it contains information about all paths of length 0 or 1, which have no vertices in their interiors.

The computation of $C_{ij}^{(k)}$ from $\boldsymbol{C}^{(k-1)}$ is given by $C_{ij}^{(k)} = C_{i,k}^{(k-1)} \wedge C_{k,j}^{(k-1)} \vee C_{i,j}^{(k-1)}$. This equation says that there is a path from i to j having only interior vertices numbered $\leq k$ if there is a path from i to j having only interior vertices numbered $\leq k-1$ or if there is a path from i to k having only interior vertices numbered $\leq k-1$ and a path from k to j having only interior vertices numbered $\leq k-1$.

Example

$$A = \begin{pmatrix} 0 & 0 & 0 & 0 & 1 \\ 0 & 0 & 1 & 0 & 0 \\ 1 & 0 & 0 & 0 & 0 \\ 0 & 0 & 1 & 0 & 0 \\ 0 & 1 & 0 & 0 & 0 \end{pmatrix} \qquad I = \begin{pmatrix} 1 & 0 & 0 & 0 & 0 \\ 0 & 1 & 0 & 0 & 0 \\ 0 & 0 & 1 & 0 & 0 \\ 0 & 0 & 0 & 1 & 0 \\ 0 & 0 & 0 & 0 & 1 \end{pmatrix}$$

$$C^{(0)} = A \vee I$$

$$C^{(1)} = \begin{pmatrix} 1 & 0 & 0 & 0 & 1 \\ 0 & 1 & 1 & 0 & 0 \\ 1 & 0 & 1 & 0 & 1 \\ 0 & 0 & 1 & 1 & 0 \\ 0 & 1 & 0 & 0 & 1 \end{pmatrix} \qquad C^{(2)} = \begin{pmatrix} 1 & 0 & 0 & 0 & 1 \\ 0 & 1 & 1 & 0 & 0 \\ 1 & 0 & 1 & 0 & 1 \\ 0 & 0 & 1 & 1 & 0 \\ 0 & 1 & 1 & 0 & 1 \end{pmatrix}$$

$$C^{(3)} = \begin{pmatrix} 1 & 0 & 0 & 0 & 1 \\ 1 & 1 & 1 & 0 & 1 \\ 1 & 0 & 1 & 0 & 1 \\ 1 & 0 & 1 & 1 & 1 \\ 1 & 1 & 1 & 0 & 1 \end{pmatrix} \qquad C^{(4)} = \begin{pmatrix} 1 & 0 & 0 & 0 & 1 \\ 1 & 1 & 1 & 0 & 1 \\ 1 & 0 & 1 & 0 & 1 \\ 1 & 0 & 1 & 1 & 1 \\ 1 & 1 & 1 & 0 & 1 \end{pmatrix}$$

$$C^{(5)} = \begin{pmatrix} 1 & 1 & 1 & 0 & 1 \\ 1 & 1 & 1 & 0 & 1 \\ 1 & 1 & 1 & 0 & 1 \\ 1 & 1 & 1 & 1 & 1 \\ 1 & 1 & 1 & 0 & 1 \end{pmatrix} \qquad \#$$

The connectivity matrix is readily seen to be the same as the matrix of the reflexive-transitive closure of the relation described by the adjacency matrix.

D. LABELED GRAPHS

Many graphs contain information in addition to the set of vertices and the edges connecting them. Examples of such graphs are flowcharts in which vertices or edges are labeled, scheduling charts in which edges are labeled with operations and the required time to accomplish the operations, distance charts in which vertices represent termini and edges are labeled with distances, and graphs of finite state machines (to be developed in Section G).

In this section, we introduce the basic definition of a labeled graph; in the following sections we show how defining an appropriate computational algebra for the graph allows different functions to be described concisely as algorithms in that algebra.

If G is a directed graph, we often label the edges of G as shown in Figure 2-5.

Definition

A labeled graph, G_L, is a directed graph, G, together with a labeling relation $R' \subseteq (V \cup E) \times L$, where L is the set of labels. (In many but not all cases, R' is a function. Generally, R' may be viewed as a function that maps edges or nodes to the set of labels.)

Example

For the graph of Figure 2-5, there is an underlying graph $G = (V = \{1, 2\}, E = \{(1, 1), (1, 2), (2, 1)\})$, and relation $R = \left\{ \dfrac{(1, 1)}{a}, \dfrac{(1, 1)}{b}, \dfrac{(1, 2)}{c}, \dfrac{(1, 2)}{b}, \dfrac{(2, 1)}{a}, \dfrac{1}{A'}, \dfrac{2}{B} \right\}.$ #

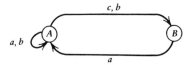

Figure 2-5. A labeled graph.

Figure 2-6

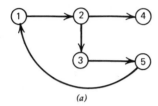

(a)

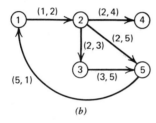

(b)

One can also label the edges in such a way as to compute the information on the cycles of a graph, where a *cycle* is a path of length ≥ 1. The labeling, which can be used to compute cycles (or paths), labels an edge from a to b: "(a, b)."

Example

In Figure 2-6a a directed graph is shown. Figure 2-6b shows this graph as a labeled graph in which the sequence of labels along any path is a description of that path.   #

E. EXTENSIONS OF FUNCTIONS AND RELATIONS

In the preceding section, we described the labeled, directed graph as an underlying graph, G, with a relation, R. The relation R immediately gives us information on all labeled paths of length 1, where a labeled path of length k is interpreted as a sequence of labels encountered in a path of length k. Our objective is to develop an algorithmic process for calculating all possible sequences of labels along paths of any length. The appropriate concept is the extension of the relation R from edges to paths of length greater than 1. We consider first the extension of functions from elements to sets.

Definition

Let $f: S \to T$ and let $f_E: P(S) \to P(T)$ be the extension of f to subsets of S defined by

$$f_E(S') = \{y \mid y = f(x) \text{ for some } x \text{ in } S'\}$$

It is clear that for unit sets f_E acts "the same as" f (except that while f maps elements to elements, f_E maps unit sets to unit sets in this case).

Example

Let s be the successor function, $s(n) = n + 1$, defined on the natural numbers. Then s_E is the extension of s, and

$$s_E(\{1\}) = \{2\}$$
$$s_E(\{2, 4, 6\}) = \{3, 5, 7\} \quad \#$$

Similarly, we may extend a function on a set, S, to a function on sequences of elements taken from S.

Definition

Let $f: S \to T$ and let $s' = (s_0, s_1, s_2, \ldots)$ be a sequence of elements from S; the extension of f to sequences is defined by

$$f_E(s') = (f(s_0), f(s_1), f(s_2), \ldots)$$

Example

$f: a \to A, b \to B, A \to A, B \to B$ can be extended to sequences as $f_E(abABa) = ABABA$. $\#$

Similarly, we may extend f to sets of sequences, and

$$f_E(\{aba, bAA, bABA\}) = \{f_E(aba), f_E(bAA), f_E(bABA)\}$$
$$= \{ABA, BAA, BABA\}$$

Also, $f: A \times B \to T$ may be extended to a function on sets of elements; $f_E: P(A) \times P(B) \to P(T)$ by $f_E: (A', B') \to \{t \mid t = f((a, b))$ for some a in A', b in $B'\}$.

Example

Concatenation of sets of strings is the extension of the operation of concatenation of strings. Thus

$$^\circ\{a, bc, de\} \cdot \{af, gec\} = \{aaf, agec, bcaf, bcgec, deaf, degec\} \quad \#$$

The extension of relations proceeds in the same way; recall only that relations may be viewed as functions on the power set.

Example

If $R = \left\{\dfrac{a}{b}, \dfrac{a}{c}, \dfrac{b}{b}, \dfrac{c}{a}, \dfrac{c}{b}\right\}$, then $R_E = \left\{\dfrac{\{a\}}{\{b, c\}}, \dfrac{\{b\}}{\{b\}}, \dfrac{\{c\}}{\{a, b\}}, \dfrac{\{a, b\}}{\{b, c\}},\right.$

$\left.\dfrac{\{a, c\}}{\{a, b, c\}}, \dfrac{\{b, c\}}{\{a, b\}}, \dfrac{\{a, b, c\}}{\{a, b, c\}}\right\}$.   #

F. CALCULATING THE SET OF LABELED PATHS

We now develop an algorithmic procedure for calculating the set of labeled paths in a directed, labeled graph.

Definition

Let the matrix M_G be the matrix of a (relational) directed graph, and let L be a labeling function of G; then M_L, the matrix of the labeled graph, is defined by $M_{L_{i,j}} = x$, where x is the set of labels of the ijth branch of G.

Example

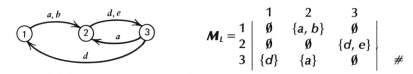

$$M_L = \begin{array}{c|ccc} & 1 & 2 & 3 \\ \hline 1 & \emptyset & \{a, b\} & \emptyset \\ 2 & \emptyset & \emptyset & \{d, e\} \\ 3 & \{d\} & \{a\} & \emptyset \end{array}$$ #

The set of labeled paths is $(M_L)^+ = \overset{\infty}{\underset{i=1}{\cup}} (M_L)^i$, where $(M_L)^1 = M_L$ and $(M_L)^{k+1} = (M_L)^k \circ M_L$, the binary operation $\circ$ on matrices is $(M_L)^k \circ M_L = (M_L)^k \underset{C}{\cup} M_L$, $\cup$ is set union, C is concatenation extended to sets of strings, and we write

$$\emptyset\, C\, x = x\, C\, \emptyset = \emptyset$$

$$a\, C\, b = ab$$

$$\{a, b\}\, C\, \{d, e\} = \{ad, bd, ae, be\}$$

Example

$$M_L = \begin{bmatrix} \emptyset & \{a, b\} & \emptyset \\ \emptyset & \emptyset & \{d, e\} \\ \{d\} & \{a\} & \emptyset \end{bmatrix}$$

$$M_L \circ M_L = \begin{bmatrix} \emptyset & \emptyset & \{a, b\}\, C\, \{d, e\} \\ \{d, e\}\, C\, \{d\} & \{d, e\}\, C\, \{a\} & \emptyset \\ \emptyset & \{d\}\, C\, \{a, b\} & \{a\}\, C\, \{d, e\} \end{bmatrix}$$

$$= \begin{bmatrix} \emptyset & \emptyset & \{ad, ae, bd, be\} \\ \{dd, ed\} & \{da, ea\} & \emptyset \\ \emptyset & \{da, db\} & \{ad, ae\} \end{bmatrix} \quad \#$$

Example

The graph of Figure 2-6*b* has the matrix:

$$M_L = \begin{pmatrix} \emptyset & \{(1, 2)\} & \emptyset & \emptyset & \emptyset \\ \emptyset & \emptyset & \{(2, 3)\} & \{(2, 4)\} & \{(2, 5)\} \\ \emptyset & \emptyset & \emptyset & \emptyset & \{(3, 5)\} \\ \emptyset & \emptyset & \emptyset & \emptyset & \emptyset \\ \{(5, 1)\} & \emptyset & \emptyset & \emptyset & \emptyset \end{pmatrix}$$

With the matrix operation $\circ$ being $\underset{C}{\cup}$, we verify that $M_L \circ M_L = (M_L)^2$ is:

$$(M_L)^2 = \begin{pmatrix} \emptyset & \emptyset & \{(1, 2)(2, 3)\} & \{(1, 2)(2, 4)\} & \{(1, 2)(2, 5)\} \\ \{(2, 5)(5, 1)\} & \emptyset & \emptyset & \emptyset & \{(2, 3)(3, 5)\} \\ \{(3, 5)(5, 1)\} & \emptyset & \emptyset & \emptyset & \emptyset \\ \emptyset & \emptyset & \emptyset & \emptyset & \emptyset \\ \emptyset & \{(5, 1)(1, 2)\} & \emptyset & \emptyset & \emptyset \end{pmatrix}$$

It is easy to check that $(M_L)^2$ describes all paths of length 2 in Figure 2-6*b*.   #

The computation of distances in a graph can be carried out in a similar way if the appropriate operations are chosen for the matrix product. In computing the minimum distance in a path of length k or less, the appropriate operations are $\left(\begin{matrix} \min \\ + \end{matrix} \right)$ to add the partial paths and choose the minimum among all paths. If G is an undirected graph, with an edge of distance between vertices a and b, there are edges (a, b) and (b, a), each labeled *l*, in the corresponding directed graph G'.

Example

Figure 2-7a shows an undirected graph of distances. The distances between vertices are shown in M_D, where the distance between any vertex and itself is 0, the distance between distinct vertices a and b is the label shown on the edge from a to b if

Figure 2-7

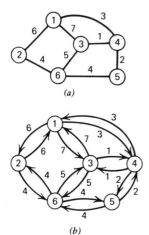

(a)

(b)

there is an edge, and the distance from vertex *a* to vertex *b* is ∞ if there is no edge from *a* to *b*.

$$M_D = \begin{pmatrix} 0 & 6 & 7 & 3 & \infty & \infty \\ 6 & 0 & \infty & \infty & \infty & 4 \\ 7 & \infty & 0 & 1 & \infty & 5 \\ 3 & \infty & 1 & 0 & 2 & \infty \\ \infty & \infty & \infty & 2 & 0 & 4 \\ \infty & 4 & 5 & \infty & 4 & 0 \end{pmatrix}$$

The minimum distance along two or fewer edges is given by

$$M_D \overset{min}{\underset{+}{}} M_D = (M_D)^2.$$

$$(M_D)^2 = \begin{pmatrix} 0 & 6 & 4 & 3 & 5 & 10 \\ 6 & 0 & 9 & 9 & 8 & 4 \\ 4 & 9 & 0 & 1 & 3 & 5 \\ 3 & 9 & 1 & 0 & 2 & 6 \\ 5 & 8 & 3 & 2 & 0 & 4 \\ 10 & 4 & 5 & 6 & 4 & 0 \end{pmatrix} \#$$

G. FINITE STATE MACHINES—APPLICATIONS

In Chapter 1 we defined a finite state machine as an abstract model consisting of a finite state set and a finite input alphabet. Each input symbol caused a transition from state to state. The machine was assumed to start in a given initial state and, in response to a finite input sequence, move through a sequence of states. After the sequence of inputs the machine is in some state, and an output function β determines whether the sequence is accepted, $\beta = 1$, or rejected, $\beta = 0$.

Let M be a finite state machine. M consists of a set of states, S, an initial state s_0, an alphabet, Σ, and a pair of functions $\delta: \Sigma \times S \to S$, and $\beta: S \to \{0, 1\}$. We may represent M as a graph G using the following rules. The vertices of G are labeled with states S of M, by a $1-1$ onto function; thus, there is exactly one vertex in G for each state in M. There is an edge from a vertex labeled s_i to a vertex labeled s_j, just in case there is some σ in Σ with $\delta(\sigma, s_i) = s_j$. An edge from s_i to s_j is labeled with $\{x \mid \delta(x, s_i) = s_j\}$. The vertex corresponding to the initial state, s_0, has an arrow leading into it. (Note that this is just a convention for denoting the initial state; it is not an edge of the graph.) The mapping, β, is shown by providing a double circle for the vertices labeled by states s for which $\beta(s) = 1$.

Example

$\quad M = (S, \Sigma, s_0, \delta, \beta)$, $\Sigma = \{a, b, c\}$, and $S = \{s_0, s_1, s_2\}$.

δ	a	b	c		β	
s_0	s_0	s_1	s_2		s_0	0
s_1	s_2	s_1	s_1		s_1	1
s_2	s_1	s_0	s_2		s_2	0

The graph, G, corresponding to M is:

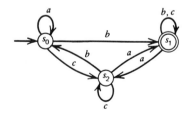

(Note that in drawing the graph of the finite state machine, we have conformed to the accepted practice of omitting set brackets on the labeling of the edges.)   #

The advantage of such an abstract model is that it can be

used to describe a variety of different concrete situations, just as the finite state machine has. In compilers the finite state model is used as a lexical analyzer for scanning the input stream and generating tokens. In such a lexical scanner, the input alphabet is the stream of symbols in the source code, and the states are often represented as locations in the program text of the lexical scanner. In problem-solving situations, the input alphabet is the set of problem actions or allowable moves, and the states are representations of intermediate situations.

Example

In a hypothetical programming language, an identifier is defined as any sequence of letters and digits, ending in a space, whose initial symbol is a letter. Letting the input set be $A = \{l, d, s, n\}$, where l = letter, d = digit, s = space, and n = none of the above, a machine to accept valid identifiers is:

$$M = (S, A, \delta, s_0, \beta):$$

δ	l	d	s	n		β	
s_0	s_1	s_3	s_0	s_3	s_0	0	
s_1	s_1	s_1	s_2	s_3	s_1	0	
s_2	s_1	s_3	s_0	s_3	s_2	1	
s_3	s_3	s_3	s_3	s_3	s_3	0	#

Example

The finite state machine of the preceding example is represented by the graph, G.

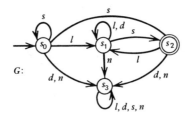

G:

#

Example

This example illustrates the use of a finite state machine and its graph representation in modeling a problem-solving situation. In a certain land there is a tribe of cannibals who are visited by missionaries. There are two missionaries and two cannibals on the left bank of the river. They wish to cross the river, but they have a rowboat that can be rowed by a single person—missionary or cannibal—and can carry only two people. Thus a sequence of moves will be necessary. For obvious reasons, one cannot allow more cannibals than missionaries at any place if there is at least one missionary at that place. The set of finite states that exist in the problem are the descriptions of the number of missionaries on the left bank, (0, 1, or 2), the number of cannibals on the left bank, (0, 1, or 2), and the bank on which the boat is located (left or right).

The graph of the machine for this problem has the actions as the input alphabet. Each action can be described as an ordered pair (m, n), where m is the number of missionaries in the boat and n is the number of cannibals in the boat.   #

Notation. A state will be denoted by a sequence of three symbols, s_1, s_2, and s_3, where s_1 is the number of missionaries on the left, s_2 is the number of cannibals on the left, and s_3 is the boat location.

Example (cont.)

The starting state is $2, 2, L$ and the only accepting state is $0, 0, R$. The graph of the finite state machine is G. A dead state is used for all nonacceptable actions from which recovery is impossible.

A portion of the graph showing the starting state and the states connected to it is:

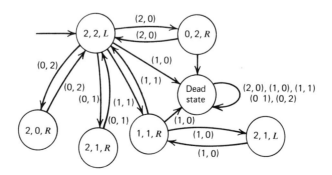

A path in the graph from the initial state to the final state is:

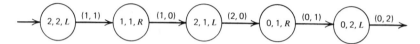

In a problem-solving situation where the number of states is large (17 in this case) and the number of actions is large (5 in this case), a program can be written to construct the transition function of the machine from the problem description. #

H. EQUIVALENCE RELATIONS ON GRAPHS

Consider the problem of designing a finite state machine to check whether n is divisible by 3 from the binary representation $B = b_1 b_2 \ldots b_k$ of a natural number, n. The inputs to the machine are successive bits of B: if we scan B from left to right, the sequence of inputs is $b_1, b_2, \ldots b_k$; if we scan B from right to left, the sequence of inputs is $b_k, b_{k-1}, \ldots, b_1$. We will first design a finite state machine in which the input is the left-to-right scan sequence and then design a finite state machine in which the input is the right-to-left scan sequence.

Left-to-Right Input Sequence

A natural assignment of states is to have three states, S_0, S_1, and S_2. Each state represents a different integer remainder: $S_0 -$ remainder $= 0$; $S_1 -$ remainder $= 1$; and $S_2 -$ remainder $= 2$. If the input sequence processed is $b_1, \ldots, b_i$, the machine will be in the state corresponding to the integer whose binary representation is $b_1, b_2, \ldots, b_i$. If $(b_1, b_2, \ldots, b_i)_2 \equiv q \pmod 3$, the machine will be in state S_q. Then, if b_{i+1} is the next input, the state after $b_1, b_2, \ldots, b_{i+1}$ will be S_r, where $(b_1, b_2, \ldots, b_i, b_{i+1})_2 \equiv r \pmod 3$. The transition from S_q to S_r can be determined by finding the dependence of r on q; it is easy to verify that $r \equiv 2 \cdot q + b_{i+1} \pmod 3$, since concatenating b_{i+1} shifts the first i digits left and thus multiplies the value of the first i digits by 2. Based on the equation relation q and r, the following transition table can be computed.

	0	1
S_0	S_0	S_1
S_1	S_2	S_0
S_2	S_1	S_2

Since the empty sequence corresponds to the integer 0, S_0 is the start state. The final state, representing the integers divisible by 3, is also S_0. The complete finite state machine is represented by the following graph.

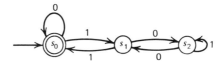

Example
   $(110101)_2 = 53$. $b_1 = 1$, $b_2 = 1$, $b_3 = 0$, $b_4 = 1$, $b_5 = 0$, and $b_6 = 1$. It can be verified from the finite state machine that the sequence of states visited in processing this number is S_0, S_1, S_0, S_0, S_1, S_2, S_2. The numbers represented by the initial positions of the input are:

$$(\lambda)_2 = \quad 0 \equiv 0 (\mathrm{mod}\ 3)$$
$$(1)_2 = \quad 1 \equiv 1 (\mathrm{mod}\ 3)$$
$$(11)_2 = \quad 3 \equiv 0 (\mathrm{mod}\ 3)$$
$$(110)_2 = \quad 6 \equiv 0 (\mathrm{mod}\ 3)$$
$$(1101)_2 = 13 \equiv 1 (\mathrm{mod}\ 3)$$
$$(11010)_2 = 26 \equiv 2 (\mathrm{mod}\ 3)$$
$$(110101)_2 = 53 \equiv 2 (\mathrm{mod}\ 3) \quad \#$$

Right-to-Left Input Sequence
   If we try the same kind of state assignment in right-to-left processing, we find that the next state depends on whether we have processed an odd or even number of digits. Thus, if

$$(b_i, b_{i+1}, \ldots, b_k)_2 \equiv q (\mathrm{mod}\ 3)$$

then

$$(b_{i-1}, b_i b_{i+1}, \ldots, b_k) \equiv q + b_{i-1} (\mathrm{mod}\ 3) \qquad \text{if } k - i + 1 \text{ is even}$$
$$\equiv q + 2 \cdot b_{i-1} (\mathrm{mod}\ 3) \quad \text{if } k - i + 1 \text{ is odd}$$

   In order to process the input from right to left, there is an apparent need to remember whether the last input processed was in an even position or an odd position. Based on this analysis, the following state transition diagram can be developed (using the symbols v for even and d for odd).

	0	1
$q_{0,v}$	$q_{0,d}$	$q_{1,d}$
$q_{0,d}$	$q_{0,v}$	$q_{2,v}$
$q_{1,v}$	$q_{1,d}$	$q_{2,d}$
$q_{1,d}$	$q_{1,v}$	$q_{0,v}$
$q_{2,v}$	$q_{2,d}$	$q_{0,d}$
$q_{2,d}$	$q_{2,v}$	$q_{1,v}$

With initial state $q_{0,v}$ and final states $q_{0,v}$ and $q_{0,d}$, the following finite state machine results.

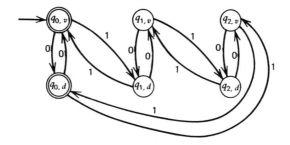

Example

$(110101)_2 = 53$.

The sequence of states visited in processing this number from right to left is:

$$q_{0,v}, \ q_{1,d}, \ q_{1,v}, \ q_{2,d}, \ q_{2,v}, \ q_{0,d}, \ q_{2,v}$$

The numbers represented by initial positions of the input are:

$$(\lambda)_2 = \quad 0 \equiv 0 (\text{mod } 3)$$
$$(1)_2 = \quad 1 \equiv 1 (\text{mod } 3)$$
$$(01)_2 = \quad 1 \equiv 1 (\text{mod } 3)$$
$$(101)_2 = \quad 5 \equiv 2 (\text{mod } 3)$$
$$(0101)_2 = \quad 5 \equiv 2 (\text{mod } 3)$$
$$(10101)_2 = 21 \equiv 0 (\text{mod } 3)$$
$$(110101)_2 = 53 \equiv 2 (\text{mod } 3) \quad \#$$

Notice that more states are used to describe the right-to-left processing than are used to describe the left-to-right processing. This suggests that perhaps some of the states in our second

machine are redundant—a state S is redundant if there is another state S' that serves the same purpose (i.e., S is redundant if
S is equivalent to, but not identical to, S'). The appropriate
definition of equivalence of states S and S' follows.

Definition
    Two states S and S' are equivalent *if every sequence σ
starting in S gives the same output as every sequence σ starting
in S'; that is,*

$$\beta(\delta_\sigma(S)) = \beta(\delta_\sigma(S'))$$

for each input sequence σ, including the null sequence.

An alternative definition that can be shown by induction
follows.

Definition (Alternative)
    Two states S and S' are equivalent *if $\beta(S) = \beta(S')$ and if for
each input symbol, x, $\delta_x(S)$ is equivalent to $\delta_x(S')$.*

Computation of the equivalence relation on the vertices of a
finite state machine is not covered here. For the machine that
processes inputs from right to left, the following partition into
equivalence classes can be verified.

$$\{\{q_{0,v},\ q_{0,d}\},\ \{q_{1,v},\ q_{2,d}\},\ \{q_{2,d},\ q_{2,v}\}\}$$

If we replace each equivalence class by a single state that
represents the equivalence class and provide appropriate tran-
sitions, we can develop a more economical machine that
computes the same function on input sequences.

Definition
    Let $G = (V, E)$ *be a relational directed graph and let R be an
equivalence relation on the set of vertices of G. G/R is the graph
obtained from G by treating each set of nodes in a block of the
partition corresponding to R as a single node and, whenever
there is a path between a vertex in block b_i and b_j, there is a path
between corresponding nodes. G/R is said to be a quotient
graph.*

Figure 2-8

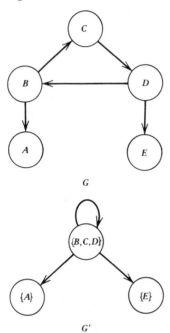

G

G'

Example

See Figure 2-8.   #

We define the relation *aRb* if there is a path from *x* to *y* and a path from *y* to *x*, or if *x* = *y*. Then it is easy to see that *R* is an equivalence relation on the set of vertices. A pair of graphs *G* and *G'* = *G*/*R* are shown in Figure 2.8.

If G_L is a labeled graph, there is an underlying graph, *G*, and a labeling function, *f*. Then the labeled graph may also be factored by an equivalence relation, as follows.

1.   *G*/*R* is the underlying graph, as defined.

2.   Noting that each edge in *G*/*R* corresponds to a set of edges in *G*, the labeling function for the quotient graph is the extension to sets of the labeling function for *G*.

Applying this reduction procedure to the right-to-left finite state processor yields the left-to-right finite state machine shown earlier in this section (with the modified labels).

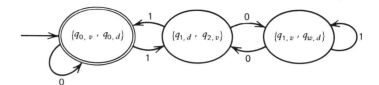

REVIEW

Directed, labeled graphs bridge the gap between the kinds of processing and input that people do efficiently and the form of computation that is carried out best by computers. Directed, unlabeled graphs correspond to relations, so the matrix representation of finite relations can be used to represent finite graphs. The graph point of view is also useful in developing the insights that can lead to more efficient algorithms. An example of a more efficient algorithm suggested by the graph representation is Warshall's algorithm for the computation of the transitive closure of a finite relation.

Labeled graphs can be used to represent finite state machines or other problem situations. Operations on such graphs can be described by an appropriate algebra operating on the matrices of the labeled graphs.

PROBLEMS

P   1.   Let $G_1 = (V_1, E_1)$, $G_2 = (V_2, E_2)$ be digraphs. Define $G_1 \times G_2 = (V_3, E_3)$, where $V_3 = V_1 \times V_2$, $E_3 \subseteq V_3 \times V_3$, and $E_3 = \{(x, y) \mid x = (a, b), y = (c, d), (a, c) \in E_1, \text{ and } (b, d) \in E_2\}$.

   (a) Let G_1 be

   Show the graph $G_1 \times G_2$.

   (b) Show that if G_1 and G_2 are graphs of posets, so is $G_1 \times G_2$.

E   2. Let G be

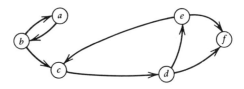

(a) Describe G as a relation.

(b) Show the graph of the transitive closure of G.

(c) If R is a relation, let R_E be the smallest reflexive relation containing R. Then R^*, the reflexive, transitive closure of R, is defined by $R^* \underset{\Delta}{\equiv} (R_E)^+ = (R^+)_E$. Show the reflexive, transitive closure of G.

P   3. Define "subgraph" and check that your definitions have the following properties.

(a) If G is a subgraph of G' and G' is a subgraph of G'', then G is a subgraph of G''.

(b) If G is a subgraph of G' and G' is a subgraph of G, then $G' = G$.

P   4. Let G be a directed graph with no cycles and let R be the relation corresponding to G. For any pair of vertices, a, b, let R_1 be the relation defined as follows: $(a, b) \in R_1$ if $a = b$ or if there is a path from a to b.

(a) Show that R_1 is a partial order.

(b) What conditions must G satisfy if R_1 is an equivalence relation?

(c) Describe an algorithm for computing R_1, given R.

E   5. Let G be the following directed graph.

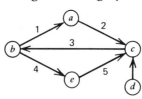

(a) Give the standard description of G.

(b) Give two cycles in G.

(c) Give two paths in G.

E 6. (a) For a graph G shown in the following figure, give the relation corresponding to G, R_G.

(b) Show the matrix representation of R_G.
(c) Calculate R_{G^*}, the reflexive-transitive closure of R_G.
(d) Show the relational graph of R_{G^*}.

E 7. Draw the graph of the relation "less than" on $\{1, 2, 3, 4\}$.

E 8. (a) Describe flowcharts as labeled graphs.
(b) Using the description of part a, give a description of

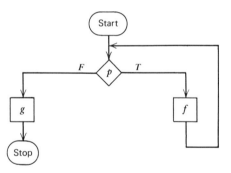

as a labeled, directed graph.

P 9. A directed graph is said to be ordered if at each vertex there is an ordering on the edges that leave that vertex. Using your flowchart model (Problem 8), explain why such an ordering may be desirable.

P 10. (a) Compute $\{1, 2, 3, 4\} \theta \{1, 2, 3, 4\}$ where θ is $+$, and where θ is $\times$.
(b) Write a program to compute the extension of an operation from elements to sets. (Recall that a set has no repeated elements.)

E 11. $R = \left\{\frac{a}{b}, \frac{b}{b}, \frac{b}{c}, \frac{a}{a}\right\}.$

   (a) Compute the transitive closure of R.
   (b) Show the graph of R.
   (c) Show the graph of the transitive closure of R.

P 12. Let G be a directed graph and R the reflexive closure of
       the relation $\hat{R}$, where $\hat{R}$ is defined so that $a\hat{R}b$ if and only
       if a and b are contained in any cycle of the graph. Show
       that R is an equivalence relation. Explain why $\hat{R}$ may not
       be an equivalence relation.

E 13. Let G be

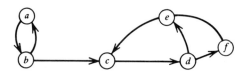

       Describe G as a relation. Compute R as described in
       Problem 12. Compute G/R.

P 14. Given a finite state machine, M, let s_i, s_j be a pair of
       states. We have s_iRs_j if, for every sequence σ, the
       sequence σ leads from s_i to a final state if and only if σ
       leads from s_j to a final state.
       (a) Prove that R is an equivalence relation.
       (b) M is defined by

δ	a	b		β	
1	2	6		1	0
2	5	6		2	1
3	5	4		3	1
4	3	6		4	0
5	5	1		5	0
6	2	6		6	0

$s_0 = 1$

       Draw the graph of M.
       (c) What is δ_{aab}?
       (d) The following relation, T, is given to you on the
       states of M: $R = \{(1, 1), (2, 2), (2, 3), (3, 2), (3, 3), (4, 6),$
       $(6, 4), (4, 4), (6, 6), (5, 5)\}$. In particular, 2 and 3 are

equivalent, as are 4 and 6. You may verify both of these. Draw the graph of G/R.

(e) In the machine corresponding to G/R, what is δ_{aab}?

E 15. Consider an 8-gallon bucket, a 5-gallon bucket, and a 3-gallon bucket. The 8-gallon bucket is initially full. The others are initially empty. A move consists of pouring from a source bucket to a destination bucket until the destination bucket is full or the source bucket is empty. Develop a graph solution to find a sequence of moves that can lead to the 8-gallon and 5-gallon buckets containing 4 gallons each.

P 16. Let G be an undirected graph. G is said to have a Euler path if there is a closed path using all edges and traversing each edge exactly once.

(a) Show that a connected graph has a Euler path if and only if the number of vertices having an odd number of incident edges is zero or two.

(b) Apply your reasoning from part a to find a Euler path for the following graph.

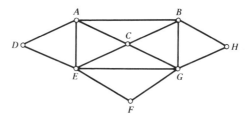

E 17. A Euler cycle is a Euler path (Problem 16) that is a cycle. Find a Euler cycle for the following graph.

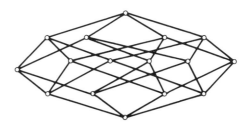

P 18.   *Definition.   A Hamilton path in a graph is a path in an undirected graph that starts from some vertex i, ends at a vertex j, never passes through either vertex i or j, and passes through every other vertex exactly once.*
Show a connected graph of more than two vertices that has no Hamilton path.

* 19.   A Hamilton cycle is a Hamilton path that is a cycle. Show that the following graph has no Hamilton cycle.

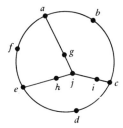

P 20.   (a) Write a computer program to compute a Euler path for a graph.
(b) Write a computer program to compute a Hamilton path for a graph.

E 21.   Consider the problem of the farmer with a dog, a goat, and a cabbage. The farmer must cross the river and can carry only one additional "passenger." If the goat and the cabbage are ever left alone, the goat will eat the cabbage. Draw a graph for this problem, with nodes labeled (x, y) where $x \leq \{$dog, cabbage, goat$\}$ and $y = $ left or $y = $ right. x should represent the objects on the left bank of the river and y the position of the farmer. Your graph should include all valid combinations of (x, y) as node labels and directed edges from (x_1, y_1) to (x_2, y_2) whenever it is possible to go from the situation represented by (x_1, y_1) to the situation represented by (x_2, y_2) in one river crossing; for example,

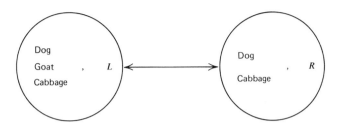

(By finding a path in the graph from

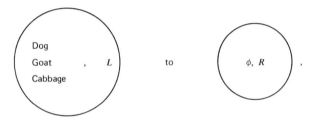

you can solve the farmer's problem of how to transport all the passengers safely.)

P 22. Show that the two definitions in the text of equivalent state in a finite state machine are the same; that is, states S_1 and S_2 are equivalent under the first definition if and only if they are equivalent under the second definition.

SUGGESTED PROGRAMMING EXERCISES

Problem 20 is a programming problem. So are the following exercises.

1. Program Warshall's algorithm.

2. Program the finite state machine to check identifiers, as given in Section G.

3. Program the missionaries and the cannibals problem for three missionaries and three cannibals. (Do not construct the machine; let your program do the work.)

4. Write a program to solve Problem 15.

3

algebraic systems

SUMMARY

Algebraic systems are introduced explicitly in this chapter. An algebraic system consists of a set of objects and a set of rules for operating on those objects. To computer scientists, the most familiar example of such a system is a set of data and a set of instructions for transforming these data.

The algebraic systems that we consider are highly idealized versions of such computational systems for at least two reasons. First, an analysis of a complicated algebraic system would necessitate very cumbersome notation, which would obscure the basic thought patterns that we are trying to clarify; therefore, we do not try to include either partial algebras or heterogeneous algebras, even though both of these generalizations are clearly relevant to computer science. Our motives here are pedagogic. Second, a detailed analysis of a particular computational system would be of further use only to those whose special interest is in that computational system. In a theoretical approach, it is better not to commit to any particular system, but to use an eclectic approach.

After defining the notion of an algebraic system, some of the simplest algebraic systems are introduced, and a discussion of the application of these simpler algebraic concepts in the development of programs follows.

The principle idea of interest to computer scientists in algebraic systems is the relationship between different algebraic systems, which is introduced in this chapter and is applied here and in subsequent chapters. Informally, each algebraic system represents a certain point of view, and a transformation of the algebra is a shift of vantage point. Thus some aspect of an algebraic system, A, may be more simply described by an algebraic system, B, where B is derived from A by a homomorphism, a rule for transcribing algebraic systems. For example, simple harmonic motion in one dimension is a homomorphic image of uniform circular motion in two dimensions.

The idea of a homomorphism is a very powerful concept that formalizes the idea of translation, which is developed in later chapters. But we can also see that a suitable collection of homomorphisms can be used to represent a more complex structure by a set of simpler structures, much as an architect's different views of a building (plan, elevation, etc.) cumulatively

Figure 3-1. Generalizing some ideas from sets to algebras.

Sets:	Equivalence relation	Partition	Assignment of an element to an equivalence class
Algebras:	Congruence	Admissible partition	Homomorphism

substitute for a model of the building. In this way, homomorphisms are central to any theory of decompositions.

The importance of the homomorphism in the study of algebraic systems motivates a presentation of this idea from several aspects, so the relationship to congruences and to admissible partitions is developed. This relationship is also useful because it builds on the development, in Chapter 1, of the connection between equivalence relations and partitions, and is an illustration of the principle of generalization and analogy (see Figure 3-1).

In the concluding sections of this chapter, one more class of algebraic systems, groups, is introduced. Group theory is an optional topic at this level. The best known application of group theory in computer science—the Krohn-Rhodes decomposition theory—is well beyond the student, as is probably Polya's counting theorem. Applications of group theory in coding theory are also available. However, some group theory is briefly offered here simply because it is an elegant illustration of the formalism and demonstrates how all the algebraic notions carry forward as more algebraic structure is introduced.

A. ALGEBRAIC SYSTEMS

Definition

An algebraic system (or, more briefly, an algebra), A, consists of a set of elements, C, called the carrier, a set of operators, Ω, and a mapping, E, called the effect. Each operator θ in Ω is assigned a fixed natural number, called the rank, which establishes the number of operands required by the operator. For any fixed integer, k, the set of k-ary operators in Ω is denoted Ω_k, and $\Omega = \bigcup_{k=0}^{\infty} \Omega_k$. E is a mapping assigning an n-ary function, $C^n \to C$, to each n-ary operator in Ω.

Example

Let $C = $ the set of integers, and $\Omega = \Omega_2 = \{+, \times, -\}$. Then (C, Ω, E) is an algebraic system, where E assigns to each opera-tor its usual meaning. The effect of the binary operator, $+$, is that it takes a pair of operands and yields a unique result. Thus $E(+) = \{((x, y), z| z$ is the sum of x and y$\}$, $E(-) = \{((x, y), z)| z$ is x minus y$\}$, and $E(*) = \{((x, y), z)| z$ is the product of x and y$\}$. If $((x, y), z)$ is an element of $E(\theta)$, we usually write $\theta(x, y) = z$. Thus $+(3, 4) = 7$, $-(5, 2) = 3$, and $*(4, 7) = 28$. In particular, if θ is a binary operator, we often use the *infix notation* and write $x\theta y$ instead of $\theta(x, y)$.   #

Example

Let $C = $ the set of integers, $\Omega_1 = \{\bar{\ }\}$, $\Omega_2 = \{+, -\}$, and $\Omega = \Omega_1 \cup \Omega_2 \cdot \Omega_1$, the set of unary operations, contains only the unary negation operator, $\bar{\ }$, where $\bar{5}$ means "negative 5." Since our definition requires that each operator have a fixed rank, we need different symbols for subtraction, $-$, which is a binary operator, and for the unary operator, $\bar{\ }$.   #

We also require that the effect of each operator is a *total function*. The following example shows an *incorrect* definition of an algebraic system.

Example

$C = $ the set of natural numbers. $\Omega = \Omega_2 = \{-\}$. $E(-) = \{((x, y), z)| z$ is x minus y, and $x \geq y\}$. In this case, $-(4, 5)$ is undefined, so that (C, Ω, E) is not an algebraic system by our definition.   #

The ranking function that assigns unique ranks to operators allows a parenthesis-free notation to be used for terms in the algebra, as follows.

1.   If x is in C, then x is a term.

2.   For each w in Ω_k and k terms $x_1, \ldots, x_k$, the expression $wx_1x_2 \ldots x_k$ is a term.

Example

$C = \{0, 1, 2, \ldots\}$ and $\Omega = \Omega_2 = \{+, x\}$; then the following are terms (selected randomly): $\{0, 1, 2, \ldots, +00, +01, x\,01, \ldots, x + 011, \ldots, +1 + x\,010, \ldots\}$. These terms have a more familiar form in infix notation: $+ + 1\,x\,023$ is written $1 + 0\,x\,2 + 3$.   #

Operators of rank 0 denote elements of the carrier. Note that by the definition of terms, these 0-ary operators are themselves terms.

Example

$\Omega_0 = \{0, 1\}$, $\Omega_2 = \{+\}$, and $\Omega = \Omega_0 \cup \Omega_2$. The set of terms generated by these operators have values ranging over the natural numbers if the effect of $+$ is taken to be addition.   #

Algebraic systems have been generalized so that the effect of an operator may be a partial function, and the resulting systems are called *partial algebras*. Another generalization allows different sets of carriers yielding heterogeneous algebras. In a heterogeneous algebra, the rank of an operator is a sequence of carrier types.

Example

The formulation of a finite state machine as a heterogeneous algebra is natural since we have two carriers, $C_1 =$ set of states and $C_2 =$ set of inputs. δ, the next state function, is an operator with operands in $C_1 \times C_2$, and the effect of δ is a mapping from $C_1 \times C_2$ to C_1.   #

In the rest of this chapter, the usual practice of not explicitly stating the effect E will be used. Each operator, θ, in the operator set Ω will in each case be interpreted as a unique operation of the appropriate rank, and the notation for the algebraic system $A = (A, \Omega)$ will be used.

B. GROUPOIDS

Definition

If A is an algebraic system and $\Omega = \{\theta\}$ and θ is of rank 2, then A is called a groupoid.

A groupoid consists of the carrier set, A, and a single binary operator, θ. The effect of θ is a total function $\theta: A^2 \to A$, and we say that θ is *closed* on A.

Examples

Let N be the natural numbers; then $(N, \{+\})$ and $(N, \{\times\})$ are groupoids, since the sum and product of a pair of natural numbers are natural numbers. $(N, \{-\})$ is not a groupoid, because subtraction is not closed on the natural numbers. If we consider the integers, Z, then $(Z, \{-\})$ is a groupoid because we have closure.   #

Definition

If A is a groupoid and θ is associative, A is called a semigroup.

Examples

As before, $(N, \{+\})$ and $(N, \{\times\})$ are semigroups, since addition and multiplication are associative. On the other hand, $(Z, \{-\})$ is not a semigroup, since subtraction is not associative. There are many important examples of semigroups in computer science.

(a) The semigroup whose carrier is the set of all relations on a set A, and whose operation is composition. Note that the set is closed (the composition of relations is a relation) and associative $[(R_1 \circ R_2) \circ R_3 = R_1 \circ (R_2 \circ R_3)]$.

(b) The semigroup whose carrier is the set of all functions on a set A and whose operation is composition.

(c) The semigroup over an alphabet Σ, whose carrier consists of all finite strings over Σ and whose operation is concatenation of strings.

(d) The semigroup of all functions on the finite state set of a finite state machine. The carrier consists of all functions from the state set of the machine to the state set of the machine induced by sequences of input symbols. Let δ denote the state set of the machine and let $|s|$ denote its cardinality. Then the carrier cannot have more than $|s|^{|s|}$ elements, and will usually have fewer elements.   #

If A is an algebraic system with carrier A and operator set $\Omega = \{\theta, i\}$, and $(A, \{\theta\})$ is a semigroup and i is a nullary operator [i.e., having rank 0] with $\theta(x, i) = \theta(i, x) = x$ for each x in A, then i is called an *identity* and A is a *monoid*. Thus a monoid is a semigroup with an identity element.

Examples

In the semigroup of finite strings over an alphabet, Σ, the identity element is the string of length zero, called the null string, λ. Denoting the concatenation of x and y, $x \cdot y$, we have $x \cdot \lambda = \lambda \cdot x = x$ for any string x.

In the semigroup of relations over a set S the identity element is the identity relation, $\{(x, x) \mid x \in S\}$.   #

Given a groupoid with a finite carrier, the operation is usually specified by means of an operations table, similar to the familiar multiplication tables (see Chapter 1). The condition for associativity is that $(a\theta b)\theta c = a\theta(b\theta c)$ for each triple of elements a, b, and c chosen from the carrier, which is readily programmed.

C. APPLICATION OF ALGEBRA TO THE CONTROL STRUCTURE OF A PROGRAM

If G is a groupoid, its operator may have special properties such as associativity, or certain of its elements may have special properties such as identity.

Examples

G_1:	θ_1	a	b	c		G_2:	θ_2	a	b	c
	a	a	b	c			a	a	a	a
	b	a	b	c			b	b	b	b
	c	a	b	c			c	c	c	c

In G_1, each of the elements acts as an identity on the left: $a\theta x = x$, $b\theta x = x$, $c\theta x = x$. In G_1 each of the elements acts as a zero on the right: $x\theta a = a$, $x\theta b = b$, $x\theta c = c$. [*Note.* The name "zero" is derived from the analogous zero element of the groupoid $(N, \{*\})$.]

In G_2, on the other hand, there are several right identities and several left zeros.   #

Fact. If a groupoid has both a left zero and a right zero, the same unique element is both left zero and right zero. If a groupoid has both a left identity and a right identity, the same unique element is both a left and right identity.

Proof. Suppose R is a right zero and L is a left zero in a groupoid, $(G, \{\theta\})$; then $L = L\theta R$, since L is a left zero, while $L\theta R = R$, since R is a right zero and, therefore, $R = L$.

Suppose r is a right identity and l is a left identity. Then $r = l\theta r$, since l is a left identity, and $l\theta r = l$, since r is a right identity and, therefore, $l = r$.   #

Example

$$G: \quad \begin{array}{c|cc} \theta & a & b \\ \hline a & a & b \\ b & b & b \end{array}$$

Here G is a monoid with an identity element, a, and a zero element, b.   #

Let $A_i 1 \leq i \leq n$ be an array of truth values; we wish to compute the conjunction of the A_i's. The following two programs will do.

$$p = \bigwedge_{i=1}^{n} A_i$$

PROGRAM 1

```
condition ← TRUE
do i = 1 to N
  condition ← condition ∧ Aᵢ
end
```

PROGRAM 2

```
condition ← TRUE
while condition do i = 1 to N
  condition ← condition ∧ Aᵢ
end
```

Program 1 runs through the whole linear array, A, to compute p. Program 2, however, makes use of the additional fact about the algebraic operation, $\wedge$; that is, it has a zero. The important point is that program 2 can be understood as a variation of program 1, which is applicable whenever the algebra has a zero.

Groupoid of logical values and $\wedge$:

$$G = (\{\text{TRUE, FALSE}\}, \{\wedge\})$$

The operation table is:

∧	TRUE	FALSE
TRUE	TRUE	FALSE
FALSE	FALSE	FALSE

The algebra of the groupoid is generally relevant to the evaluation of the sequence $(S_1, S_2, \ldots, S_n)$ over the groupoid $(G, \{\theta\})$ by the computation of $(\ldots ((S_1 \theta S_2) \theta S_3) \ldots S_n)$. The following programs are applicable.

PROGRAM 3.   Most general.

$$\nu \leftarrow S_1$$
$$i \leftarrow 2$$
$$\underline{\text{while }} i \leq n \underline{\text{ do }} \underline{\text{begin}}$$
$$\qquad \nu \leftarrow \nu\theta \, S_i$$
$$\qquad i \leftarrow i + 1$$
$$\qquad \underline{\text{end}}$$

PROGRAM 4.   The groupoid has a left identity.

$$\nu \leftarrow 1$$
$$i \leftarrow 1$$
$$\underline{\text{while }} i \leq n \underline{\text{ do }} \underline{\text{begin}}$$
$$\qquad \nu \leftarrow \nu\theta \, S_i$$
$$\qquad i \leftarrow i + 1$$
$$\qquad \underline{\text{end}}$$

PROGRAM 5.   The groupoid has a left zero.

$$\nu \leftarrow S_1$$
$$i \leftarrow 2$$
$$\underline{\text{while }} i \leq n \underline{\text{ and }} \nu \neq 0 \underline{\text{ do }} \underline{\text{begin}}$$
$$\qquad \nu \leftarrow \nu\theta \, S_i$$
$$\qquad i \leftarrow i + 1$$
$$\qquad \underline{\text{end}}$$

PROGRAM 6.   The groupoid has a left zero and a left identity.

$$\nu \leftarrow 1$$
$$i \leftarrow 1$$
while $i \leqq n$ **and** $\nu \neq 0$ **do begin**
$\quad\quad \nu \leftarrow \nu\theta\, S_i$
$\quad\quad i \leftarrow i+1$
$\quad$ **end**

PROGRAM 7.   The groupoid has a left zero and a left identity and programs 1 and 2 can now be described as the evaluation of a sequence over the groupoid $G = (\{\text{TRUE, FALSE}\}, \{\wedge\})$. If S_i, $1 \leqq i \leqq n$, is the array of truth values, the most general program is:

$$\nu \leftarrow S_1$$
$$i \leftarrow 2$$
while $i \leqq n$ **do begin**
$\quad\quad \nu \leftarrow \nu \wedge S_i$
$\quad\quad i \leftarrow i+1$
$\quad$ **end**

Recognition of both the presence of a left zero and left identity and no zero divisors allows the program to be written as

$$\nu \leftarrow 1$$
$$i \leftarrow 1$$
while $i \leqq n$ **do begin**
$\quad\quad$ **if** $S_i = 0$ **then do begin**
$\quad\quad \nu \leftarrow 0$
$\quad\quad$ **exit**
$\quad\quad$ **end**
$\quad \nu \leftarrow \nu \wedge S_i$
$\quad i \leftarrow i+1$
end

Note.   If the groupoid may have zero divisors, the test must be done on ν, as in program 6.

Many more variations are possible if the groupoid is associative or commutative or if other special properties arise such as:

$i \leftarrow 1$
while $(i < n$ and $S_i = \text{TRUE})$
    $i \leftarrow i + 1$
    end
$v \leftarrow S_i$

which can be used for the iterative evaluation over the groupoid of truth values and $\wedge$.

If x and y are not zeros of the groupoid, but z is a zero, and if $x\theta y = Z$, then x and y are said to be *zero divisors*.

Example

$$G: \quad \begin{array}{c|cccc} x_4 & 0 & 1 & 2 & 3 \\ \hline 0 & 0 & 0 & 0 & 0 \\ 1 & 0 & 1 & 2 & 3 \\ 2 & 0 & 2 & 0 & 2 \\ 3 & 0 & 3 & 2 & 1 \end{array}$$

In the groupoid $G = (\{0, 1, 2, 3\}, \{x_4\})$ of multiplication, modulo 4, 2 is a zero divisor, since $2 \times 2 = 4 = 0 \bmod 4$. This monoid has a zero, 0, and an identity 1.   #

D. HOMOMORPHISMS

A homomorphism is a way of "carrying over" operations from one algebraic system to another.

Example

Let $A = (A, \{\text{composition}\})$ be the algebra of Algol programs, and let $P = (P, \{\text{composition}\})$ be the algebra of PDP-11 programs. If we have a homomorphism from A to P, we can translate programs from Algol to PDP-11 code. This particular homomorphism is what is known as syntax-directed translation.   #

Definition

A homomorphism ϕ of a groupoid $A = (A, \{\theta\})$ into a groupoid $B = (B, \{\times\})$ is a mapping $\phi: A \rightarrow B$ such that $\phi(a_1 \theta a_2) = \phi(a_1) \times \phi(a_2)$.

Figure 3-2.   Diagram representing homomorphisms.

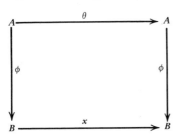

The definition of homomorphism maps the carrier of one algebra to the carrier of the second algebra. However, only mappings that carry the operations across from the first to the second algebra are homomorphisms.

The diagram of the concept of a homomorphism (Figure 3-2) is often helpful.

Starting at the upper left corner with elements of A, we may operate on them in the groupoid, A, symbolized by the top horizontal arrow, and carry results over to B through the right vertical arrow. Alternatively, we may take the starting operands in A, map these to corresponding elements of B, symbolized by the left vertical arrow, and operate on them in B through the lower horizontal arrow. The fact that ϕ is homomorphism is a consequence of the (implicit) fact that whichever path (horizontal and then vertical or vertical and then horizontal) is taken from the upper left corner of the diagram to the lower right corner of the diagram, the same result is obtained. Figure 3-2 also *commutes*, which means that the result is independent of the path.

Examples

(a) $G =$ (positive reals, $\{\times\}$), $H =$ (reals, $\{+\}$), ϕ: positive reals $\rightarrow$ reals, ϕ: $x \mapsto \log x$. The fact that the logarithm function is a homomorphism from the positive reals under multiplication to the reals under addition allows us to multiply by adding logarithms using the homomorphism equation $\log(x \times y) = \log(x) + \log(y)$. The diagram is:

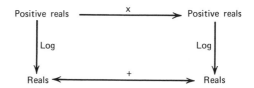

(b) $G = (\{\text{TRUE, FALSE}\}, \{\wedge\})$, $H = (\{\text{TRUE, FALSE}\}, \{\vee\})$, $\phi: \{\text{TRUE, FALSE}\} \to \{\text{TRUE, FALSE}\}$, $\phi: x \mapsto x$. This homomorphism is one of DeMorgan's laws, $\rceil(x \wedge y) = \rceil x \vee \rceil y$, which we diagram as:

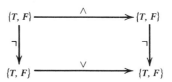

Let A, B, and C be three groupoids where $A = (A, \{\theta_1\})$, $B = (B, \{\theta_2\})$, and $C = (C, \{\theta_3\})$; suppose that ϕ_1 is a homomorphism from A to B and ϕ_2 is a homomorphism from B to C. It is easy to see that $\phi_3 = \phi_1 \circ \phi_2$ is a homomorphism from A to C.

Let x and y be any elements in A; since ϕ_1 is a homomorphism,

$$\phi_1(x\theta_1 y) = \phi_1(x)\theta_2\phi_1(y) \tag{3-1}$$

Now $\phi_1(x)$ and $\phi_1(y)$ are elements in B and, since ϕ_2 is a homomorphism,

$$\phi_2(\phi_1(x)\theta_2\phi_1(y)) = \phi_2(\phi_1(x))\theta_3\phi_2(\phi_1(y)) \tag{3-2}$$

Substituting Equation 3-1 into Equation 3-2,

$$\phi_2(\phi_1(x\theta_1 y)) = \phi_2(\phi_1(x))\theta_3\phi_2(\phi_1(y)) \tag{3-3}$$

But $\phi_2(\phi_1(x))$ is just the result of applying ϕ_1 and then ϕ_2; that is, $\phi_2(\phi_1(x)) = (\phi_1 \circ \phi_2)(x)$. Equation 3-3 may be rewritten as

$$(\phi_1 \circ \phi_2)(x\theta_1 y) = (\phi_1 \circ \phi_2)(x)\theta_3(\phi_1 \circ \phi_2)(y) \tag{3-3a}$$

Figure 3-3.   Composition of homomorphisms.

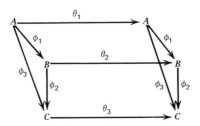

Since $\phi_1 \circ \phi_2 = \phi_3$, this becomes

$$\phi_3(x\theta_1 y) = \phi_3(x)\theta_3\phi_3(y) \tag{3-4}$$

showing that ϕ_3 is a homomorphism from A to C.

The composition of homomorphisms can also be diagrammed as in Figure 3-2.

E. AN EXAMPLE OF HOMOMORPHISMS APPLIED

The idea of a homomorphism is to select some part of the information in an algebraic system and to use this partial information. The particular homomorphism, Q, used here will map letters of the alphabet onto a set of two symbols, $\{-, x\}$

$$\text{where } Q: \alpha \mapsto - \qquad \text{if } \alpha \text{ is a vowel}$$
$$\alpha \mapsto x \qquad \text{if } \alpha \text{ is a consonant}$$

This homomorphism will be developed in the process of solving the following cryptogram[1] to provide further clues in the solution.

ZF CZA BFTRFB ZGB MAOVSTD

CFHH ZEB VA VFFX AY EVMFBSATB

The following steps show the reasoning.

1.   The pattern of the ninth word, VFFX, suggests that F is a vowel, either E or O.

---

[1]A cryptogram is a puzzle obtained by using a permutation π of the alphabet to represent a message, with each letter α replaced by $\pi(\alpha)$. Solving the cryptogram means computing π^{-1}.

2.   Since VFFX has a double vowel in its interior, V and X must be consonants, leading to a partial vowel-consonant picture.

$$ZF\ CZA\ BFTRFB\ ZGB\ MAOVSTD$$
$$CFHH\ ZEB\ VA\ VFFX\ AY\ EVMFBSATB$$

[Here as in the rest of this section the vowel-consonant indicators $\{-, x\}$ will be shown below the corresponding letters.]

Now C and H are probably consonants because CFHH, and Z is a consonant because of ZF. This leads to

$$ZF\ CZA\ BFTRFB\ ZGB\ MAOVSTD$$
$$CFHH\ ZEB\ VA\ VFFX\ AY\ EVMFBSATB$$

3.   Since V is a consonant in VA, A is a vowel. Since A occurs as the second letter of a two-letter word, VA, and as the first letter of a two-letter word, AY, it is probably O, and from step 1, the F is E, which leads to

$$\begin{array}{l}E\quad O\ E\quad E\qquad\qquad O\\ ZF\ CZA\ BFTRFB\ ZGB\ MAOVSTD\end{array}$$

$$\begin{array}{l}E\qquad\quad O\ EE\ O\qquad E\ O\\ CFHH\ ZEB\ VA\ VFFX\ AY\ EVMFBSATB\end{array}$$

(The vowel-consonant information is recorded below the cryptogram and the specific letters $\pi^{-1}(\alpha)$ are recorded above the letters of the cryptogram when π^{-1} is known.]

4.   B is probably a consonant because of BFTRFB; G is a vowel because of ZGB, and E is a vowel because of ZEB. This leads to

$$\begin{array}{l}E\quad O\ E\ E\qquad\qquad O\\ ZF\ CZA\ BFTRFB\ ZGB\ MAOVSTD\end{array}$$

$$\begin{array}{l}E\qquad\quad O\ EE\qquad\quad E\ O\\ CFHH\ ZEB\ VA\ VFFX\ AY\ EVMFBSATB\end{array}$$

5.   A judicious guess based on the pattern of the first two words is that they are HE WHO; since this gives $\overset{\text{WE}}{\text{CFHH}}$, we conclude

also that $\pi^{-1}(H) = L$, giving

HE WHO E   E H   O
ZF CZA BFTRFB ZGB MAOVSTD

WELL H   O EE O   E   O
CFHH ZFB VA VFFX AY EVMFBSATB

6.   A further guess from the pattern $\underset{x-x}{VA} \underset{x-x}{VFFX} \underset{-x}{AY}$ (with O EE O above) is that $\pi^{-1}(V) =$ N, $\pi^{-1}(X) = D$, and $\pi^{-1}(Y) = F$, giving:

HE WHO E   E H   O N
ZF CZA BFTRFB ZGB MAOVSTD

WELL H   NO NEED OF N E   O
CFHH ZEB VA VFFX AY EVMFBSATB

7.   The next guess is that $\pi^{-1}(B) = S$ and $\pi^{-1}(\{G, E\}) = \{I, A\}$, because of $\underset{x-x}{ZEB}$, $\underset{x-x}{ZGB}$ (with H S above and H S above) and the fact that E and O are already chosen ($\pi(E) = F$, $\pi(O) = A$). [An alternative guess might have been $\pi^{-1}(B) = D$; nondeterministically one could try to pursue both guesses in parallel, or one can backtrack.] This gives

HE WHO SE   ES H S   O N
ZF CZA   BFTRFB ZGB MAOVSTD

WELL H S NO NEED OF N ES O S
CFHH ZEB VA VFFX AY EVMFBSATB

8.   Most likely, $\pi^{-1}(G) = I$ and $\pi^{-1}(E) = A$, and the vowel-consonant pattern suggests that at least $\pi^{-1}(\{S, T, D\})$ contains a vowel.

Exercise.   Complete the calculation of the cryptogram solution.

F. CONGRUENCES

In discussing partitions and equivalence relations (Chapter 1), we observed that partitions and equivalence relations are two ways of describing the same thing: whenever we are given a partition, we can define a corresponding equivalence relation,

Figure 3-4.   The correspondence between set notions and their algebraic counterparts.

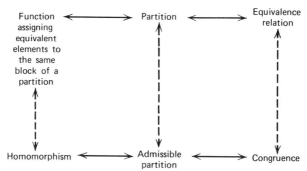

and vice versa. In the same way, the notions of congruence, admissible partition, and homomorphism are different representations of the same concept. A congruence is just an equivalence relation that "works" in an algebraic context, and an admissible partition is just a partition on whose blocks the algebra "works." Of course, these vague ideas will be made precise in this section and the next section.

With this point of view, a homomorphism can be thought of as a function that assigns equivalent elements to the same block of the admissible partition. Forgetting about the algebra, the corresponding set concept is a function that, given an equivalence relation, assigns equivalent functions to the same block of a partition. The analogy can be described as in Figure 3-4.

In an algebraic system we often have equivalent elements. We can regard two subroutines as equivalent if they can be used interchangeably, even though the internal structure of the subroutines may be quite different. We have also seen that in finite state machines, different sequences of inputs induce the same state transformation and thus act as equivalent input sequences. An equivalence relation in an algebraic system is called a *congruence*.

Definition

An equivalence relation E over the elements of the carrier of a groupoid $(A, \{\theta\})$ is called a congruence if xEy and zEw imply $x\theta z \, E \, y\theta w$ for every x, y, z, and w in A. If x is congruent to y, we write $x \equiv y$.

The notion of a congruence is that an element can serve as an alternate for any equivalent element, because they are equivalent, and yield equivalent results.

Examples

(a) In the groupoid of the integers under addition, x and y are said to be congruent mod k, $x \equiv y$ (mod k) if x and y have the same remainder when divided by k. We see that congruence mod k is an equivalence relation. (Why?) Now $x \equiv y$ (mod k) implies that if $x = p_1 k + q_1$, then $y = p_2 k + q_1$. Likewise, $z \equiv w$ (mod k) implies that $z = p_3 k + q_2$ and $w = p_4 k + q_2$. Therefore, $x + z = p_1 k + q_1 + p_3 k + q_2 = (p_1 + p_3)k + q_1 + q_2$, and $y + w = p_2 k + q_1 + p_4 k + q_2 = (p_2 + p_4)k + q_1 + q_2$. Thus $x + z$ and $y + w$ have the same remainder when divided by k, and "congruence mod k" is indeed preserved by the groupoid operation.

(b) In the groupoid of the integers under multiplication, congruence mod k is again shown to be a congruence. Using x, y, z, and w as defined in Example a,

$$x * z = (p_1 k + q_1) * (p_3 k + q_2) = (p_1 p_3 k + q_1 p_3 + p_1 q_2)k + q_1 q_2$$

and

$$y * w = (p_2 k + q_1) * (p_4 k + q_2) = (p_2 p_4 k + q_1 p_4 + p_2 q_2)k + q_1 q_2$$

so that $x * z \equiv y * w$ (mod k) is indeed a congruence.

(c) In Example a, choosing $k = 2$, we have the following familiar rules.

$$\text{even} + \text{odd} = \text{odd} + \text{even} = \text{odd}$$

$$\text{odd} + \text{odd} = \text{even} + \text{even} = \text{even}$$

In our language, calling the evenness or oddness of an integer its *parity*, we can say that parity is a congruence for the groupoid of the integers under addition. You may also verify that it is a congruence for the groupoid of the integers under multiplication.   #

The definition of congruence generalizes to an arbitrary algebraic system where we require that for each operator θ of rank n_θ and each set of operands $\{x_i | 1 \le i \le n_\theta\}$, the proposition

$$x_i E y_i, \ 1 \le i \le n_\theta$$

implies

$$\theta(x_1, x_2, \ldots, x_n) E \theta(y_1, y_2, \ldots, y_n)$$

G. ADMISSIBLE PARTITIONS

A congruence is an equivalence relation that is preserved in an algebraic system in that equivalent operands yield equivalent results. We have seen that each equivalence relation has a partition associated with it; each block of the partition consists of the set of all elements equivalent to one of the elements of the block. Conversely, for each partition, there is an equivalence relation where two elements are equivalent if and only if they are in the same block of the partition. The class of partitions associated with congruences is called admissible partitions. In the finite case, we can test a partition to see if it is admissible by computing the corresponding equivalence relation and then verifying that it is a congruence. Alternatively, we can test the partition directly. We will give the details for the case of a groupoid, $A = (A, \{\theta\})$.

Let $\pi = \{B_1; B_2; \ldots ; B_n\}$ be a partition of A. The product of blocks $B_i \theta B_j$ in the groupoid $\{B_i, \{\theta\}\}$ is given by the extension of θ to sets

$$B_i \theta B_j = \{x \theta y \,|\, x \in B_i, y \in B_j\}$$

Definition

A partition $\pi = \{B_1; \ldots ; B_n\}$ of the carrier of a groupoid $A = (A, \{\theta\})$ is called admissible *if, for any pair of blocks, B_i and B_j, there is a block B_k such that the product of B_i and B_j is contained in B_k; that is, $B_i \theta B_j \subseteq B_k$.*

Example

Let A be the groupoid of the natural numbers under addition. The following partition is admissible.

$$\pi = \{B_1; B_2\}$$

where

$$B_1 = \{0, 2, 4, \ldots\}$$
$$B_2 = \{1, 3, 5, \ldots\}$$

We can check that π is admissible, since

$$B_1 + B_1 = \{0, 2, 4, \ldots\} = B_1$$
$$B_1 + B_2 = B_2 + B_1 = \{1, 3, 5, \ldots\} = B_2$$

and

$$B_2 + B_2 = \{2, 4, 6, \ldots\} \subseteq B_1$$

(Note that this example also shows that the groupoid operation extended to a pair of blocks is only guaranteed to be contained in a third block, but that it is often *not* equal to that block.) #

Example

Again we have the groupoid of natural numbers under addition. Now consider the partition $\pi = \{B_1; B_2\}$, where

$$B_1 = \{x \mid x \text{ is a perfect square}\} = \{0, 1, 4, 9, 16, \ldots\}$$
$$B_2 = \{2, 3, 5, 6, 7, 8, 10, \ldots\}$$

then

$$B_1 + B_1 = \{0, 1, 2, 4, \ldots\}$$

Since $\{1, 2\} \subseteq B_1 + B_1$, we see that $B_1 + B_1$ is not contained in either B_1 or B_2. Consequently, the partition into squares and nonsquares is not admissible. #

If E is an equivalence relation on the carrier of a groupoid, $A = (A, \{\theta\})$, and E is a congruence, the partition of A into congruence classes of E is admissible, and these classes form a groupoid with the operation of the groupoid being the extension to sets of the original groupoid operation. This new groupoid is called a *quotient groupoid* and is denoted A/E. In general, the algebra of congruence classes is called *quotient algebra* or *factor algebra*.

H. RELATIONSHIP BETWEEN HOMOMORPHISMS, CONGRUENCES, AND ADMISSIBLE PARTITIONS

Fact. A congruence and an admissible partition can be associated with every homomorphism, and vice versa.

Let $G = (G, \{\theta\})$ be a groupoid, and suppose that E is a congruence over G. Then E is, by definition, an equivalence relation on the elements of G, and thus corresponds to a partition of the elements of G into equivalence classes. We write this partition as G/E, and observe that if B_1 and B_2 are blocks of G/E, then $B_1 \theta B_2$ will be contained in a block of G/E. That is, the partition G/E is admissible. Thus the blocks of G/E are an algebraic system whose carrier is G/E and whose binary operation is ψ, operating on blocks of elements. This groupoid $(G/E, \{\psi\})$ is denoted G/E and is called the *factor groupoid* (or *quotient groupoid*). ψ is, of course, just the operation θ on the sets of elements in the blocks, except that $B_i\theta B_j$ may be only a proper subset of B_k, and we then define $B_i\psi B_j$ to be equal to B_k.

Now let ϕ be the mapping, which takes each element in G into the block in which it resides. $x\theta y E x'\theta y'$ whenever xEx' and yEy'. Thus the equivalence class of $x\theta y$, $\phi(x\theta y)$, is the same as the equivalence class of $x'\theta y'$, $\phi(x'\theta y')$. Knowing the equivalence classes of x and y uniquely determines the equivalence class of $x\theta y$. But this is the operation on equivalence classes, ψ, and we have $\phi(x\theta y) = \phi(x)\psi\phi(y)$. If E is a congruence, the mapping ϕ of elements to their equivalence classes is a homomorphism from G to the algebra on the equivalence classes, G/E.

On the other hand, suppose that ϕ is a homomorphism from G to H. Then $E = \phi\phi^{-1}$ is a congruence on G; that is, any pair of elements that map to the same element are congruent, and a congruence class is the set of preimages of an element in H.

Figure 3-5 shows graphically the relationship between algebraic systems. If ϕ is a homomorphism from G to H, the factor groupoid, G/E, will be isomorphic to the image of G in H, which means that the algebraic systems will be the same except for renaming.

Calling the factor groupoid the *image* of the homomorphism, we have the following information.

Fact.   A homomorphic image of a monoid is a monoid.

Proof.   We must show that the image is associative and has an identity. Let the source monoid be (G, θ) and the image be (H, ψ), and let the homomorphism be ϕ.

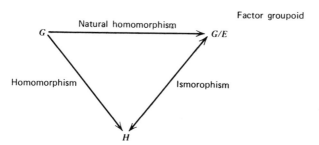

Figure 3-5.   The correspondence between homomorphisms and factor groupoids.

(a) *Identity.* Let x be an arbitrary element of G and i_G be the identity in G. $i_G\theta x = x$ and $x\theta i_G = x$, since i_G is an identity. Then $\phi(i_G\theta x) = \phi(x) = \phi(x\theta i_G)$, since ϕ is a mapping, and $\phi(i_G\theta x) = \phi(i_G)\psi\phi(x)$, since ϕ is a homomorphism, and $\phi(i_G)$ is a left identity in the image. Similarly, $\phi(x\theta i_G) = \phi(x)\psi\phi(i_G)$, and $\phi(i_G)$ is a right identity. So $\phi(i_G)$ is an identity in (H, ψ).

(b) *Associativity.* Let x, y, and z be any elements in G. $\phi(x\theta(y\theta z)) = \phi(x)\psi(\phi(y\theta z)) = \psi(x)\psi(\phi(y)\psi\phi(z))$ and $\phi((x\theta y)\theta z) = \phi(x\theta y)\psi\phi(z) = (\phi(x)\psi\phi(y))\psi\phi(z)$, but $x\theta(y\theta z) = (x\theta y)\theta z$, so $\phi(x)\psi(\phi(y)\psi\phi(z)) = (\phi(x)\psi\phi(y))\psi\phi(z)$. #

We can summarize the relationship between homomorphism and congruence relations in monoids by the *fundamental homomorphism theorem for monoids.*

If ϕ is a homomorphism from a monoid $G = (G, \{*\})$ to another monoid $H = (H, \{\cdot\})$ and E is the congruence relation $\phi \cdot \phi^{-1}$, then H is isomorphic to the quotient monoid G/E.

This fundamental homomorphism theorem is pictured in Figure 3-5.

Example of a Homomorphism

Define a relation, R', on words[2] by

$$xR'y \Leftrightarrow x = uZ_1v, \qquad y = uZ_2v$$

and Z_1, Z_2 are both in $\{\lambda, aa, bbb\}$ or both in $\{ab, bba\}$, u and v are arbitrary words over $\{a, b\}$ called the context, and R is the reflexive transitive closure of R'.

For example (underlining the contexts), $aa\underline{b}\,R\,\underline{b}$, $\underline{a}ab\,R\,\underline{a}bba$, $abba\,R\,bba\underline{ba}$, $\underline{bb}aba\,R\,\underline{bb}\,bbaa$, $bbbbaa\,R\,\underline{b}aa$.

(a) R is an equivalence relation (in fact, a congruence).

(b) There is an admissible partition.

(c) There is a homomorphism.

We now develop the homomorphism, congruence, and admissible partition for this example using two functions, ϕ and ψ, defined as follows.

(a) If xRy, then $\phi(x) = \phi(y)$, where ϕ is the remainder of the number of a's divided by 2.

---

[2]Recall that words are just sequences of letters.

(b) $\psi(x)$ is now computed. Start at the left. When $a'b'$ is encountered, if the a's seen so far are even, add 1. If the a's seen so far are odd, then subtract 1. ψ is the remainder of the number computed, after it is divided by three.

It is easy to establish the following facts.

Fact. $xRy \Rightarrow \phi(x) = \phi(y)$ and $\psi(x) = \psi(y)$.

Proof.   If x is related to y, it is related by a sequence of 0 or more substitutions from R'. But for each pair of words that are R' related, it is easy to verify that $xR'y \Leftrightarrow \phi(x) = \phi(y)$ and $\psi(x) = \psi(y)$.   #

Fact.   Every word is equivalent to one of the following:

$$\lambda, \ b, \ bb, \ a, \ ba, \ bba$$

Proof.   For any word, x, $\phi(x) \in \{0, 1\}$, and $\psi(x) \in \{0, 1, 2\}$. The following table shows the equivalent word among the specified set for each pair ϕ, ψ.

		ψ		
		0	1	2
ϕ 0		λ	b	bb
1		a	ba	bba

#

Fact. $xRy \Leftrightarrow \phi(x) = \phi(y)$ and $\psi(x) = \psi(y)$.

Fact.   Every word is equivalent to one of the following:

$$\lambda, \ b, \ bb, \ a, \ ba, \ bba$$

We call these the *normal forms* of the words.

If a word in one equivalence class is concatenated with a word in another equivalence class, the equivalence class of the result is determined by the equivalence class of the operands. Seen as a finite state machine, two words are equivalent if and only if they lead to the same state.

We can also view the equivalence class as a pair of numbers.

$$(n_b, n_a) \in \{0, 1, 2\} \times \{0, 1\}$$

n_b, n_a is the number of b's and a's in the normal form. Suppose $\Gamma(x) = (\psi(x), \phi(x))$.

$$\Gamma: x \mapsto (n_{b_1}, n_{a_1})$$
$$y \mapsto (n_{b_2}, n_{a_2})$$

Then $\Gamma: x \cdot y \vdash (n_{b_3}, n_{a_3})$, where $n_{b_3} \equiv n_{b_1} + (1 - 2n_{a_1})n_{b_2}$ (mod 3) and $n_{a_3} \equiv n_{a_1} + n_{a_2}$ (mod 2). That is, concatenation of words map into an algebra of pairs of numbers through a homomorphism.

The congruence relation for this example is R. The admissible partition is a partition where all words that lead to the same state, from the start state, are in the same block of the partition. The block of words that leads to state $(1, 1)$ is:

$$\{ba, abb, aabb, bbab, \ldots\}$$

The homomorphism Γ maps from the monoid of words over $\{a, b\}$ under concatenation to the monoid of number pairs $\{0, 1, 2\} \times \{0, 1\}$ under the operation $*$, where the operation table of $*$ is:

$*$	0, 0	1, 0	2, 0	0, 1	1, 1	2, 1
0, 0	0, 0	1, 0	2, 0	0, 1	1, 1	2, 1
1, 0	1, 0	2, 0	0, 0	1, 1	2, 1	0, 1
2, 0	2, 0	0, 0	1, 0	2, 1	0, 1	1, 1
0, 1	0, 1	2, 1	1, 1	0, 0	2, 0	1, 0
1, 1	1, 1	0, 1	2, 1	1, 0	0, 0	2, 0
2, 1	2, 1	1, 1	0, 1	2, 0	1, 0	0, 0

#

Admissible partitions and congruences are related concepts. If a congruence relation is given, the equivalence classes of that congruence are a partition of the carrier and the partition is admissible.

Conversely, if an admissible partition is available, as a partition, it determines an equivalence relation, which is a congruence.

Beware of the following, which is *not* true. A partition such that each of the members of any block is congruent to all the members of that block is not necessarily admissible. Only if the blocks are made as large as possible and include all congruent elements can you be sure that it is admissible.

Relation of Homomorphisms to Admissible Partition. If one has an admissible partition, $P = \{P_1; P_2; \ldots ; P_k\}$, the elements of the partition will determine a unique groupoid. Fixing i and j, the product

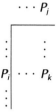

of P_i and P_j is determined. If we call this groupoid $(P, \{\theta\})$ and call the original groupoid $(G, \{\theta\})$, there is a homomorphism from $(G, \{\theta\})$ to $(P, \{\theta\})$. It carries elements of G to P, assigning to each element x of G the block of P in which x resides.

If, on the other hand, one is given a homomorphism, ϕ, mapping the carrier of one groupoid, $(G, \{+\})$, into the carrier of a second groupoid $(H, \{\times\})$, for any element k of H, the set of all the elements of G mapping to that k are a block of the partition. **Relation of Congruences to Homomorphisms.** If E is a congruence on $(G, \{+\})$, construct the admissible partition corresponding to it. This partition determines a homomorphism.

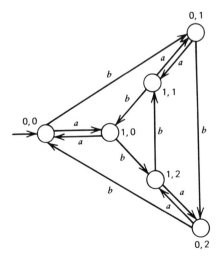

Figure 3-6.   A finite state machine for the congruence of this example.

This homomorphism will be from the groupoid $(G, \{+\})$ to the groupoid of the blocks of the admissible partition.

On the other hand, if ϕ is a homomorphism from $(G, \{+\})$ to $(H, \{\times\})$, this determines an admissible partition which, in turn, determines a congruence. This congruence is on the elements of G. Two elements of G are congruent if they map to the same element in H.

Example

Let x and y be words over the alphabet $\{a, b\}$. $x\rho y$ if and only if $x = uv_1w$ and $y = uv_2w$ and v_1 and v_2 are both in A or both in B, where $A = \{\lambda, aa, bbb\}$, $B = \{ab, ba\}$ and λ is the null word.

Observe that B allows us to interchange an adjacent a and b, since $u_1bav \rho u_1abv_2$. However, bba is not ρ-related to abb; although $bba \rho bab$ and $bab \rho abb$, ρ is not transitive, as defined. To remedy this deficiency and allow us to move a's and b's freely, we let σ be the reflexive-transitive closure of ρ.

Now a sequence of applications of B allows all the a's to be collected at the beginning of a word. Any word x is σ-related to y where y is a string of a's followed by a string of b's. Then A allows us to reduce the number of a's to 0 or 1 and to reduce the number of b's to 0, 1, or 2.

Let $n_\alpha(w) \underset{\Delta}{=}$ the number of α's in w. Then $x\sigma y$ if and only if $|n_a(x) - n_a(y)| = 2 \cdot k$ and $|n_b(x) - n_b(y)| = 3 \cdot l$.

That σ is indeed an equivalence relation is clear, since ρ was already reflexive and symmetric. Taking the transitive closure adds transitivity, but does not change reflexity or symmetry. (Why?)

To check that σ is a congruence, you must verify that

$$x E y \text{ and } w E z = x\theta w \, E \, y\theta z.$$

In our case, θ is concatenation. But if x has a normal form $a^{n_1}b^{m_1}$,

$$n_1 \in \{0, 1\}$$
$$m_1 \in \{0, 1, 2\}$$

and if w has a normal form $a^{n_2}b^{m_2}$,

$$n_2 \in \{0, 1\}$$
$$m_2 \in \{0, 1, 2\}$$

Then $x\,w$ has a normal form $a^{n_3}b^{m_3}$, and

$$n_3 \in \{0, 1\}$$

$$m_3 \in \{0, 1, 2\}$$

obtained from the normal forms of x and w by a sequence of B moves followed by a sequence of A moves.

It is easily checked that any sequence of B moves never changes either the number of a's or the number of b's. And any sequence of A moves preserves the equivalence class. Thus, x is congruent to y in the congruence relation that we have called σ. There are six congruence classes, as follows.

$$C_1 = \{x \mid n_a(x) = 2 \cdot k, \; n_b(x) = 3 \cdot l\}$$

$$C_2 = \{x \mid n_a(x) = 2 \cdot k + 1, \; n_b(x) = 3 \cdot l\}$$

$$C_3 = \{x \mid n_a(x) = 2 \cdot k, \; n_b(x) = 3 \cdot l + 1\}$$

$$C_4 = \{x \mid n_a(x) = 2 \cdot k + 1, \; n_b(x) = 3 \cdot l + 1\}$$

$$C_5 = \{x \mid n_a(x) = 2 \cdot k, \; n_b(x) = 3 \cdot l + 2\}$$

$$C_6 = \{x \mid n_a(x) = 2 \cdot k + 1, \; n_b(x) = 3 \cdot l + 2\}$$

The admissible partition is $\{C_1; C_2; C_3; C_4; C_5; C_6\}$. You can easily verify that it is really a partition and that it is admissible. *Note that you need not go back to the congruence to do this.* It can be done directly from the admissible partition.

Suppose w_1 is a word in C_3 and w_2 is a word in C_5; then

$$n_a(w_1) = 2 \cdot k_1, \qquad n_b(w_1) = 3 \cdot l_1 + 1$$

$$n_a(w_1) = 2 \cdot k_1 \qquad n_b(w_2) = 3 \cdot l_2 + 2$$

and

$$n_a(w_1 \cdot w_2) = 2(k_1 + k_2), \qquad n_b(w_1 \cdot w_2) = 3(l_1 + l_2 + 1)$$

showing that $w_1 \cdot w_2$ is a word in C_1. Note that the congruence class of the result depended only on the congruence classes of the operands, and was independent of which particular elements were chosen.

We next show a different concrete representation of an algebraic system isomorphic to this quotient algebra; $G = (\{a, b, c, d, e, f\}, \{*\})$, where the operation table of $*$ is:

*	a	b	c	d	e	f
a	a	b	c	d	e	f
b	b	a	d	c	f	e
c	c	d	e	f	a	b
d	d	c	f	e	b	a
e	e	f	a	b	c	d
f	f	e	b	a	d	c

(Incidentally, this groupoid is a commutative group. A concrete realization of this group can be obtained as follows. Take an equilateral triangle and a line segment. Each action consists of rotating the triangle by a multiple of 120°, *and* rotating about the line segment *fixed in the triangle* through a multiple of 180°.)

POSITION OF VERTEX AFTER ACTION.

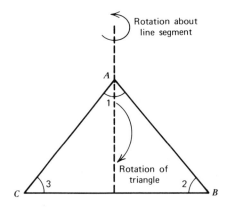

Action	1	2	3
a	A	B	C
b	A	C	B
c	B	C	A
d	B	A	C
e	C	A	B
f	C	B	A

	Triangle Motion	Line Segment Motion
a	0°	0°
b	0°	180°
c	120°	0°
d	120°	180°
e	240°	0°
f	240°	180°

There is a homomorphism from the algebra of words under concatenation just shown to the algebra of triangle motions.

Each element of: C_1 goes into a.

Each element of: C_2 goes into b.

Each element of: C_3 goes into c.

Each element of: C_4 goes into d.

Each element of: C_5 goes into e.

Each element of: C_6 goes into f.

The reason that this homomorphism works is that the groupoid of $\{C_1, C_2, C_3, C_4, C_5, C_6\}$ under concatenation is the same as the groupoid of $\{a, b, c, d, e, f\}$ under $*$, except for relabeling.   #

Some Things to Think About

1. Given a finite groupoid and a partition, P, what is the program to determine if P is admissible?

2. Given a finite groupoid and an equivalence relation, E, on the carrier of the groupoid, what is a program to determine if E is a congruence?

3. Given a mapping from the carrier of $(G, \{\theta_1\})$ to the carrier of $(H, \{\theta_2\})$, what is the program to determine if this mapping is a homomorphism?

4. For questions 1 to 3, think of an example when the following conditions are not true.
(a) A partition of the carrier of a groupoid that is not admissible.
(b) An equivalence relation on the carrier of a groupoid that is not a congruence.
(c) A mapping between groupoids that is not a homomorphism.

5. If the groupoid has an infinite carrier, it may be that many of these questions are undecidable and, therefore, algorithms to test these things are not available.

6. If $(G, \{\theta\})$ has a finite congruence, then there is a finite state machine that can separate the congruence classes.

I. PROGRAMS

Each programming language can be viewed as an algebraic system. In this algebraic system, subprograms are "composed" by given rules. For example, using Dijkstra's constructs, each elementary instruction is a program. Furthermore, if p_1 and p_2 are programs and π is a predicate, then so are:

$p_1; p_2$, which is p_1 followed by p_2

if π then p_1 else p_2

while π do p_1

If we have a programming language, L_1, with instructions $p_1, p_2, \ldots, p_k$ and predicates $\pi_1, \pi_2, \ldots, \pi_l$, we may be able to construct a second programming language, L_2, with basic instructions $q_1, q_2, \ldots, q_m$ and predicates $\zeta_1, \ldots, \zeta_n$, where we have a mapping from L_1 to L_2 that carries each instruction in L_1 to an instruction in L_2 and carries each predicate in L_1 to each predicate in L_2 in such a way that the homomorphism condition is satisfied. Then, for every program in L_1, there is an image program in L_2. This homomorphism establishes equivalence classes in L_1. Any two programs giving the same result in L_2 are equivalent in L_1.

One example of such a homomorphism is the mapping that takes each instruction into the function that it computes. These functions along with the rule of composition, may be considered a programming language. Clearly, any two programs in L_1 that compute the same function may be equivalent for many practical purposes. Indeed, the most economical of the equivalent programs is often sought.

An application of this equivalence is the set of transformations that an optimizing compiler makes.

J. GROUPS

Starting with a set and a binary operation, and introducing properties and additional operations, we have a hierarchy of algebraic systems.

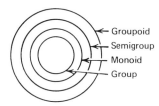

The most general system, requiring only a binary operation that is closed on the carrier, is the groupoid. If the binary operation is associative, the groupoid is a semigroup. If the semigroup has an identity element, it is a monoid. If the binary operation, θ, is invertible, the monoid is a group. θ is invertible if for each element a in the carrier there is an element a^{-1}, called the inverse of a, that satisfies the equations:

$$a\theta a^{-1} = a^{-1}\theta a = i$$

where i is the identity.

Example

(a) In the algebraic system $A = (A, \{\theta\})$, where $A = \{i, a, b\}$ and the operation table of θ is as specified, θ is invertible. The operation

θ	i	a	b
i	i	a	b
a	a	b	i
b	b	i	a

table of θ shows that $i^{-1} = i$, since $i^{-1} = i^{-1}\theta i = i\theta i^{-1} = i$; $b = a^{-1}$, since $a\theta a^{-1} = a\theta b = i$ and $a^{-1}\theta a = b\theta a = i$; and $a = b^{-1}$, since $b\theta b^{-1} = b\theta a = i$ and $b^{-1}\theta b = a\theta b = i$.

(b) In the algebraic system

$$A = (A, \{\times\}),$$

where $A = \{i, a, b\}$ and the operation table of $\times$ is

$\times$	i	a	b
i	i	a	b
a	a	i	b
b	b	b	b

,

A is a monoid, but $\times$ is not invertible, since $x \times b = b \times x = b$ for every x so that there is no b^{-1} satisfying $b^{-1} \times b = b \times b^{-1} = i$.   #

When θ is invertible, we define the unary operator, $^{-1}$ (written as a postfix exponent) which, applied to any element, yields the inverse element. We now formally define the algebraic system known as a group.

Definition

A group *is an algebraic system,* $G = (G, \Omega)$, *where* $\Omega = \{\theta, i, ^{-1}\}$ *and* θ *is a binary associative operator closed on* G; i *is a nullary operator (called the identity element of* G), *and* $^{-1}$ *is a unary operator called the inverse that satisfies the equations* $a^{-1}\theta a = a\theta a^{-1} = i$. *(We will follow the usual practice of referring to* $(G, \{\theta\})$ *as a group whenever* θ *is invertible.)*

We can deduce the following properties of inverses.

1. The inverse of the identity is the identity:

$$i\theta i^{-1} = i \qquad \text{by definition of inverse}$$
$$i\theta i^{-1} = i^{-1} \qquad \text{by definition of identity}$$

Therefore $i = i^{-1}$.

2. The inverse is unique. Suppose that a has b and c as inverses. Then $a\theta b = a\theta c$, which implies $b\theta(a\theta b) = b\theta(a\theta c)$ and, by associativity, $b\theta(a\theta b) = (b\theta a)\theta b = i\theta b = b$, while $b\theta(a\theta c) = (b\theta a)\theta c = i\theta c = c$. Therefore $b = c$.

3. The inverse of the inverse of an element a is a; that is, $(a^{-1})^{-1} = a$. Let $a^{-1} = b$. Then $a\theta b = b\theta a = i$ (definition of a^{-1}), which implies that $a = b^{-1}$ (definition of b^{-1}). Therefore, $a = (a^{-1})^{-1}$ (substituting for b).

4. $(a\theta b)^{-1} = b^{-1}\theta a^{-1}$. By associativity, $(b^{-1}\theta a^{-1})\theta(a\theta b) = b^{-1}\theta(a^{-1}\theta a)\theta b = b^{-1}\theta i\theta b = b^{-1}\theta b = i$ and, similarly, $(a\theta b)\theta(b^{-1}\theta a^{-1}) = i$.

A monoid, with binary operation θ, is said to be *solvable* if for every a, b the equations $a\theta x = b$ and $x\theta a = b$ have unique solutions.

Fact. If a monoid is solvable, it is a group.

Proof. The only requirement is that the monoid have inverses. However, since it is solvable, we can solve the equations $a\theta x = i$ and $x\theta a = i$. Let b be the solution to $a\theta x = i$, and let c be the solution to $x\theta a = i$. Then $c\theta a = i$, which implies that $(c\theta a)\theta b = i\theta b = b$. Similarly, $a\theta b = i$, which implies that $c\theta(a\theta b) = c\theta i = c$ and, by associativity, $c\theta(a\theta b) = (c\theta a)\theta b$. Thus, $b = c$ and b is a^{-1}. #

A monoid, with binary operation θ, is said to be cancellable if $a\theta x = a\theta y$ implies that $x = y$, and $x\theta a = y\theta a$ implies that $x = y$.

Fact. If a finite monoid is cancellable, it is a group; an infinite monoid is not necessarily a group if it is cancellable.

Example

The monoid of natural numbers under addition is cancellable, since $x + a = y + a$ implies $x = y$ and $a + x = a + y$ implies $x = y$. But the monoid does not have inverses. #

While discussing homomorphisms we have seen that different concrete representations of groups give apparently different algebraic systems. There is, however, a standard form of representing groups by a set of permutations under the operation of composition.

THEOREM (CAYLEY). Every finite group is isomorphic to a group of permutations (and, correspondingly, every finite monoid is isomorphic to a monoid of a set of functions on the carrier).

We illustrate Cayley's theorem by the following example.

Example

FINITE (ABSTRACT) GROUP.

θ	i	a	b
i	i	a	b
a	a	b	i
b	b	i	a

FINITE GROUP OF PERMUTATIONS.

	f_i	f_a	f_b
f_i	f_i	f_a	f_b
f_a	f_a	f_b	f_i
f_b	f_b	f_i	f_a

where $f_i = \begin{pmatrix} i & a & b \\ i & a & b \end{pmatrix}$, $f_a = \begin{pmatrix} i & a & b \\ a & a & b \end{pmatrix}$, and $f_b = \begin{pmatrix} i & a & b \\ b & b & b \end{pmatrix}$.   #

Next, we develop the group version of congruences, homomorphisms, and admissible partitions.

Definition
   A group $B = (B, \{\psi\})$ is a subgroup of $A = (A, \{\theta\})$ if B is a subset of A and ψ is the restriction of θ to B, meaning that if x and y are in B, $x\psi y = x\theta y$.
Fact. A subset T of a group $G = (G, \{\theta\})$ is a subgroup of G iff for every pair of elements a, b in T, $a\theta b^{-1}$ is also in T. [More accurately, $(T, \{\theta\})$ is the subgroup.]
Proof. If T is a subgroup, it has inverses and is closed under the operation.
   On the other hand, if a and b are in T, $a\theta a^{-1}$ is in T, so $i \in T$. Then $i\theta a^{-1} = a^{-1} \in T$, so that $a, b \in T$ implies $b^{-1} \in T$, which then implies $a\theta(b^{-1})^{-1} \in T$, so $a\theta b \in T$, and T is closed, has identity and inverses, and is a group. (The associativity of the operation carries over from the group.)   #

Definition
   If $G = (G, \{\theta\})$ is a group and $H = (H, \{\theta\})$ is a subgroup of G then, for any element $g \in G$, $H\theta\{g\}$ is called a right coset of H.

Fact. Right cosets form a partition of G.
Proof. It suffices to show that for any two cosets $H\theta\{a\}$ and $H\theta\{b\}$, either they are identical or they are disjoint, since any element x is a member of $H\theta\{x\}$.
   We claim, in fact, that if $c \in H\theta\{a\}$, then $H\theta\{a\} = H\theta\{c\}$. If $c \in H\theta\{a\}$, then $c = h\theta a$ for some $h \in H$. Then $a = h^{-1}\theta c$, since H is a subgroup, and $H\theta\{a\} = H\theta\{h^{-1}\theta c\}$ (by definition of a) =

$(H\theta\{h^{-1}\})\theta\{c\}$ (by associativity) $= H\theta\{c\}$ (since H is a subgroup $H\theta\{h^{-1}\} = H$).   #

Definition
    The order of a finite group is the cardinality of the carrier.

Corollary
    In a finite group the order of every subgroup H divides the order of the group.

    Now we have a special case of the partitions of an algebraic system, the cosets of a subgroup. In general, the partition of a group into right cosets is not admissible.

Example

θ	e	a	a^2	b	c	d
e	e	a	a^2	b	c	d
a	a	a^2	e	c	d	b
a^2	a^2	e	a	d	b	c
b	b	d	c	e	a^2	a
c	c	b	d	a	e	a^2
d	d	c	b	a^2	a	e

    We can check that $H = \{e, b\}$ gives a subgroup. The cosets of H are:

$$H\theta\{e\} = H\theta\{b\} = \{e, b\}$$
$$H\theta\{a\} = H\theta\{d\} = \{a, d\}$$
$$H\theta\{a^2\} = H\theta\{c\} = \{a^2, c\}$$

But $\{a, d\}\theta\{a^2, c\} = \{e, d, b, a\}$, so the partition into cosets of H is not admissible.   #

Definition
    A subgroup H of G is called normal if it corresponds to an admissible partition (i.e., if its right cosets form an admissible partition.)

We can define left cosets, $\{g\}\theta H$, similar to the definition of right cosets.

Fact. H is normal in G if and only if its right and left cosets coincide (i.e., $H\theta\{g\} = \{g\}\theta H$ for each g in G).

Proof. Assume that H is normal in G. Since $H\theta\{a\}$ and $\{a^{-1}\}\theta H$ are cosets, $(\{a^{-1}\}\theta H)\theta(H\theta\{a\}) = \{a^{-1}\}\theta H\theta\{a\}$ is a coset, because the coset partition is admissible. The identity is an element in $\{a^{-1}\}\theta H\theta\{a\}$, so it follows that $\{a^{-1}\}\theta H\theta\{a\} = H$ and, premultiplying by $\{a\}$, $H\theta\{a\} = \{a\}\theta H$.

On the other hand, if $H\theta\{a\} = \{a\}\theta H$, then

$$(H\theta\{a\})\theta(H\theta\{b\}) = H\theta(\{a\}\theta H)\theta\{b\}$$
$$= H\theta(H\theta\{a\})\theta\{b\}$$
$$= (H\theta H)\theta\{a\theta b\}$$
$$= H\theta\{a\theta b\}$$

and the product of cosets is a coset; that is, the coset partition is admissible. #

Let G/H denote the quotient algebra (the algebra of cosets). Then, as in the general algebraic case, $\phi: G \to G/H$ is a homomorphism, called the natural homomorphism (for the subgroup H), as in Figure 3-4.

Suppose that $\psi: G \to G_1$ is any homomorphism of groups, $G = (G, \{\theta\})$, $G_1 = (G_1, \{\theta'\})$. The preimage of the identity in G_1 forms a normal subgroup of G. First, in a homomorphism, identities map to identities, so that if $\psi(a) = i_1$ and $\psi(b) = i_1$,

$$\psi(b\theta b^{-1}) = \psi(b)\theta'\psi(b^{-1}) \qquad (\psi \text{ is a homomorphism})$$
$$= i_1\theta'\psi(b^{-1}) \qquad [\psi(b) = i_1 \text{ is given}]$$
$$= i_1 \qquad (b\theta b^{-1} \text{ is the identity in } G)$$

From these last two lines, $\psi(b^{-1}) = i_1$, and so

$$\psi(a\theta b^{-1}) = \psi(a)\theta'\psi(b^{-1})$$
$$= i_1\theta'i_1$$
$$= i_1$$

and $a\theta b^{-1}$ is a preimage of i_1. Since $a, b \in \{\text{preimages of } i_1\}$ implies $a\theta b^{-1} \in \{\text{preimages of } i_1\}$, $\{\text{preimages of } i_1\}$ form a subgroup.

That H, the set of preimages of i_1, is normal can be seen from

$$\psi(\{a^{-1}\}\theta H\theta\{a\}) = \{\psi(a^{-1})\}\theta'\psi(H)\theta'\{\psi(a)\}$$
$$= i_1\theta'i_1\theta'i_1$$
$$= i_1$$

Thus $\{a^{-1}\}\theta H\theta\{a\}\subseteq H$, but $|\{a^{-1}\}\theta H\theta\{a\}| = |H|$, so that $\{a\}\theta H = H\theta\{a\}$.

This is summarized in the following theorem.

THEOREM. To every normal subgroup H of G there corresponds a factor group G/H that is a homomorphic image of G; conversely, given a homomorphism ϕ of G onto G_1, the set of elements mapped by ϕ onto the identity of G_1 forms a normal subgroup H of G (called the kernel of ϕ).

K. GROUPS AND THEIR GRAPHS

We saw in Chapter 2 that directed graphs are a convenient way of visualizing relations. Also, we saw earlier in this chapter that graphs can represent finite state machines whose states correspond to congruence classes of the sequences over the input alphabet. In such a finite state machine the set of sequences that lead to a given state from the initial state form a congruence class.

It is, therefore, not surprising that finite groups can be represented as finite digraphs. In the graph of a group, vertices represent elements of the group and edges represent generators.

Definition

A set of generators, g, of a group $G = (G, \{\theta\})$ is a subset of the carrier, G, such that every element of G can be expressed in terms of elements of G.

Examples

(a) Let $G = (G, \{\theta\})$ be a group where $G = \{a, b, c\}$ and the operation table of θ is:

θ	a	b	c
a	a	b	c
b	b	c	a
c	c	a	b

One set of generators for this group is $g = \{b\}$, and $a = b\theta b\theta b$, $c = b\theta b$. A group that is generated by a single element is called a *cyclic* group.

(b) Let $G = (G, \{\theta\})$ be a group where G is the set of all permutations on n elements, $\{1, 2, \ldots, n\}$ and θ is the composition of permutations. A set of generators for G is p, q,

where $\qquad p = \begin{pmatrix} 1 & 2 & 3 & 4 & \cdots & n \\ 2 & 1 & 3 & 4 & & n \end{pmatrix} \qquad$ and $\qquad q =$

$\begin{pmatrix} 1 & 2 & 3 & 4 & 5 & \cdots & n-1 & n \\ 2 & 3 & 4 & 5 & 6 & & n & 1 \end{pmatrix}$. Every permutation can be

expressed as a finite product of p's and q's.

We illustrate the case of $n = 3$.

$$\pi_1 = \begin{pmatrix} 1 & 2 & 3 \\ 1 & 2 & 3 \end{pmatrix} = p\theta p$$

$$\pi_2 = \begin{pmatrix} 1 & 2 & 3 \\ 2 & 1 & 3 \end{pmatrix} = p$$

$$\pi_3 = \begin{pmatrix} 1 & 2 & 3 \\ 1 & 3 & 2 \end{pmatrix} = q\theta p$$

$$\pi_4 = \begin{pmatrix} 1 & 2 & 3 \\ 2 & 3 & 1 \end{pmatrix} = q$$

$$\pi_5 = \begin{pmatrix} 1 & 2 & 3 \\ 3 & 2 & 1 \end{pmatrix} = p\theta q$$

$$\pi_6 = \begin{pmatrix} 1 & 2 & 3 \\ 3 & 1 & 2 \end{pmatrix} = q\theta q \qquad\qquad \#$$

The graph of the group has one vertex for each element of the group, and at each vertex, v_i, for each generator, g_j, there is an edge leaving that vertex labeled with the generator, g_j, and leading to the vertex labeled with $v_i\theta g_j$. The vertex corresponding to the identity element is designated by an unlabeled, incoming edge.

Examples

(a) For the cyclic group of three elements given in the preceding example, the graph is:

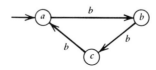

Note that this graph of the group is not unique, but depends on the choice of generator set. Choosing an alternative generator set, {c}, the corresponding graph is:

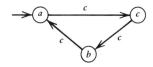

(b) For the group of all permutations of three elements under composition, the graph with the preceding generators is using the same notation.

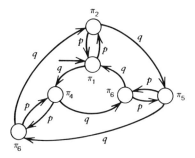

Note that in the graph of the permutation group of three elements, the expression for any element can be composed from any sequence of labels leading to the vertex corresponding to that element, and that all such expressions are equal. Thus,

$$q\theta q\theta p\theta q\theta p = p\theta q\theta p\theta q\theta q$$

since each of the sequences of labeled edges (q, q, p, q, p) and (p, q, p, q, q) lead from π_1 to π_4.   #

L. APPLICATION OF CONGRUENCES TO FINITE STATE MACHINES

Recall that a finite state machine is defined as $M = (S, \Sigma, \delta, s_0, \beta)$, where S is a finite state set, Σ is a finite input alphabet, δ is a transition function, $\delta: S \times \Sigma \to S$, $s_0 \in S$ is a start state, and β is an output function, $\beta: S \to \{0, 1\}$.

The next state function δ is extended to be a function $\hat{\delta}: S \times \Sigma^* \to S$. Thus, if $\delta(s, a) = s'$, s' is the state of the finite state machine that results when an input symbol, $a \in \Sigma$, is applied to the machine in state s. Likewise, $\hat{\delta}$ defines the state of the machine in response to an input sequence. If the machine is in state s and an input sequence w is applied, and if $\hat{\delta}(s, w) = s''$, after the sequence is applied the machine will be in state s''. Finally, if $\hat{\delta}$ is parametrized, as in Chapter 1, we have, for each input sequence w, a state mapping $\delta_w : S \to S$, where $\hat{\delta}(s, w) = \hat{\delta}_w(s)$.

The finite state machine processes strings over the input alphabet Σ, Σ^*. The set of strings over the alphabet Σ can be described, as we have seen, as a monoid under concatenation, $\Sigma^* = (\{\text{strings over } \Sigma\}, \{\cdot\})$. The identity of the monoid is λ, the null string.

Example

$M = (S, \Sigma, \delta, s_0, \beta)$, where $S = \{s_0, s_1, s_2\}$, $\Sigma = \{a, b, c\}$,

δ	a	b	c		β	
s_0	s_1	s_2	s_0		s_0	0
s_1	s_0	s_0	s_2		s_1	1
s_2	s_1	s_0	s_2		s_2	0

The graph representation of M is:

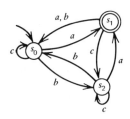

Some typical state mappings are:

$\delta_\lambda : s_0 \mapsto s_0$ $\hat{\delta}_a : s_0 \mapsto s_1$ $\hat{\delta}_b : s_0 \mapsto s_2$ $\hat{\delta}_c : s_0 \mapsto s_0$

$ s_1 \mapsto s_1$ $\phantom{\hat{\delta}_a :} s_1 \mapsto s_0$ $\phantom{\hat{\delta}_b :} s_1 \mapsto s_0$ $\phantom{\hat{\delta}_c :} s_1 \mapsto s_2$

$ s_2 \mapsto s_2$ $\phantom{\hat{\delta}_a :} s_2 \mapsto s_1$ $\phantom{\hat{\delta}_b :} s_2 \mapsto s_0$ $\phantom{\hat{\delta}_c :} s_2 \mapsto s_2$

$\hat{\delta}_{aa} : s_0 \mapsto s_0$ $\hat{\delta}_{ab} : s_0 \mapsto s_0$ $\hat{\delta}_{ac} : s_0 \mapsto s_2$ $\hat{\delta}_{ba} : s_0 \mapsto s_1$

$\phantom{\hat{\delta}_{aa} :} s_1 \mapsto s_1$ $\phantom{\hat{\delta}_{ab} :} s_1 \mapsto s_2$ $\phantom{\hat{\delta}_{ac} :} s_1 \mapsto s_0$ $\phantom{\hat{\delta}_{ba} :} s_1 \mapsto s_1$

$\phantom{\hat{\delta}_{aa} :} s_2 \mapsto s_0$ $\phantom{\hat{\delta}_{ab} :} s_2 \mapsto s_0$ $s_2 \nearrow s_2$ $\phantom{\hat{\delta}_{ba} :} s_2 \mapsto s_1$

$$\hat{\delta}_{bb}: s_0 \mapsto s_0 \qquad \hat{\delta}_{bc}: s_0 \mapsto s_2 \qquad \hat{\delta}_{ca}: s_0 \mapsto s_1 \qquad \hat{\delta}_{cb}: s_0 \mapsto s_2$$
$$s_1 \mapsto s_2 \qquad\quad s_1 \mapsto s_0 \qquad\quad s_1 \mapsto s_1 \qquad\quad s_1 \mapsto s_0$$
$$s_2 \mapsto s_2 \qquad\quad s_2 \mapsto s_0 \qquad\quad s_2 \mapsto s_1 \qquad\quad s_2 \mapsto s_0$$

$$\hat{\delta}_{cc}: s_0 \mapsto s_0 \qquad \hat{\delta}_{baa}: s_0 \mapsto s_0$$
$$s_1 \mapsto s_2 \qquad\qquad s_1 \mapsto s_0$$
$$s_2 \mapsto s_2 \qquad\qquad s_2 \mapsto s_0 \qquad\qquad \#$$

Each finite state machine defines a congruence over Σ^* in the algebraic sense. Specifically, define the equivalence relation on strings by:

$$w_1 \, E \, w_2 \qquad \text{iff } \hat{\delta}_{w_1} = \hat{\delta}_{w_2}$$

We can easily verify that E is an equivalence relation. (Do it.) Next, we can show that E is, in fact, a congruence by showing that $w_1 \, E \, w_2$ and $w_3 \, E \, w_4$ imply $w_1 \cdot w_3 \, E \, w_2 \cdot w_4$.

If $w_1 \, E \, w_2$ then $\hat{\delta}_{w_1} = \hat{\delta}_{w_2}$ (definition of E)

If $w_3 \, E \, w_4$ then $\hat{\delta}_{w_3} = \hat{\delta}_{w_4}$ (definition of E)

$\hat{\delta}_{w_1 \cdot w_3} = \hat{\delta}_{w_1} \circ \hat{\delta}_{w_3}$ (definition of $\hat{\delta}$)

$\qquad\quad = \hat{\delta}_{w_2} \circ \hat{\delta}_{w_4}$ (substituting the equalities)

$\qquad\quad = \hat{\delta}_{w_2 \cdot w_4}$ (definition of $\hat{\delta}$)

If $\hat{\delta}_{w_1 \cdot w_3} = \hat{\delta}_{w_2 \cdot w_4}$ then $w_1 \cdot w_3 \, E \, w_2 \cdot w_4$ (definition of E)

Thus E is a congruence on the set of strings over Σ, the congruence being determined by the particular finite state machine.   #

Example

In the machine M of the previous example, $ca \, E \, ba$ and $b \, E \, bc$. Thus, in any string we may replace ca by ba or by any other congruent string.   #

In fact, any finite state machine determines a congruence relation with only a finite number of congruence classes. This can easily be seen, since there are only finitely many states and, consequently, only a finite number of possible different mappings from S to S. The converse is also true. *Given any congruence relation over Σ^* with a finite number of congruence classes, there is a finite state machine defining that congruence relation.*

Example

Continuing the previous example, the complete monoid of congruence classes is:

*	λ	a	b	c	aa	ab	ac	ba	baa	bac
λ	λ	a	b	c	aa	ab	ac	ba	baa	bac
a	a	aa	ab	ac	a	ac	ab	ba	baa	bac
b	b	ba	c	b	baa	baa	bac	ba	baa	bac
c	c	ba	b	c	baa	baa	bac	ba	baa	bac
aa	aa	a	ac	ab	aa	ab	ac	ba	baa	bac
ab	ab	ba	ac	ab	baa	baa	bac	ba	baa	bac
ac	ac	ba	ab	ac	baa	baa	bac	ba	baa	bac
ba	ba	baa	baa	bac	ba	bac	baa	ba	baa	bac
baa	baa	ba	bac	baa	baa	baa	bac	ba	baa	bac
bac	bac	ba	baa	bac	baa	baa	bac	ba	baa	bac

Each string over $\{a, b, c\}$ has the same state mapping, $\hat{\delta}$, as one of the strings $\{\lambda, a, b, c, aa, ab, ac, ba, bb, baa\}$. Furthermore, given two strings whose congruence classes are known, the congruence class of the concatenation of the strings can be determined from the operation table of the congruence monoid.

Thus,

$$\hat{\delta}_{abcbca}: s_0 \mapsto s_1$$
$$s_1 \mapsto s_1$$
$$s_2 \mapsto s_1$$

can be determined directly from the finite state machine. Alternatively, using the congruence algebra, $\underline{abcb}ca\ E\ ab\underline{bc}a\ E\ \underline{abba}\ E\,ba$, where the congruence class of cb
$\quad$(1)$\qquad$(2)$\qquad$(3)

can be determined from the monoid of congruence classes as b and substituted in (1) to obtain (2). Similarly, $bc\,E\,b$ and $abba\,E\,ba$.

Note that $bc\,E\,b$ can be used in parallel processing to reduce $\underline{ab\,cbc}a$ in one step to $abba$.   #

REVIEW

Algebraic systems are abstractions of programming systems and other systems of interest to computer scientists. In this chapter, the simplest algebraic systems, groupoids, semigroups, monoids, and groups, are considered.

Particular emphasis is placed on means of relating one algebraic system to another through the related notions of homomorphism, congruence, and admissible partition.

PROBLEMS

E 1. Which of the following are algebraic systems? (In each case, use the conventional meaning of the operators, and let Z denote the integers, R the reals, X the complex numbers, and N the natural numbers.)

(a) $N, \{+, \times\}$
(b) $N, \{-, \times\}$
(c) $R, \{+, \times\}$
(d) $R, \{\times, \div\}$
(e) $Z, \{+, -\}$
(f) $Z, \{\div\}$
(g) $X, \{+, \times\}$
(h) $X - \{0\}, \{+, \times, -, \div\}$

E 2. Let $\Omega_0 = \{2\}$, $\Omega_2 = \{\times\}$, and $\Omega = \Omega_0 \cup \Omega_2$. Describe the set of values of the terms generated by Ω where the effect of "$\times$" is taken to be multiplication.

P 3. Let A and B be the permutations

$$A: 1 \to 2 \qquad B: 1 \to 2$$
$$2 \to 1 \qquad\quad 2 \to 3$$
$$3 \to 3 \qquad\quad 3 \to 1$$

Show how under composition each permutation of the set of three elements can be expressed as a product of A's and B's. (This product is generally not unique; list only one product for each element.)

P 4. The products of elements A and B in Problem 3 (e.g., $A \cdot B \cdot B \cdot A \cdot B \cdot A \cdot A \cdot B$) form congruence classes. Two products are congruent if they are the same permutation. Noting that (a) permutations are functions, and hence associative under composition, (b) $A \cdot A = B \cdot B \cdot B = I$, the identity permutation, and (c) $A \cdot B \cdot A \cdot B = I$, state an algorithmic procedure (that does not require the calculation of the composite permutations) for determining if two given products are congruent.

P 5.   Let α and β be two congruence relations on a semi-group. Show that $\alpha \cap \beta$ is also a congruence relation.

E 6.   Let M be the following finite state machine.

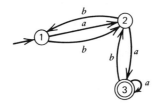

(a) Two sequences of a's and b's are equivalent if they cause the same state mapping $S \to S$. Show that this equivalence is a congruence (under concatenation of sequences).

(b) Find the shortest sequence equivalent to *abab-babbba*.

(c) Find the shortest sequence equivalent to *abb-babbaba*.

E 7.   Let x and y be words over the alphabet $\{a, b, c\}$. $x\rho y$ if and only if $x = uv_1w$ and $y = uv_2w$ and v_1 and v_2 are both in A or both in B, where $A = \{\lambda, bb, ccc\}$, $B = \{bc, cb\}$, and λ is the null word.
Example
    ab<u>cbc</u>a ρ ab<u>ccb</u>a.   #
(a) Which of the following words are ρ-related?

abc, bca, cab, cbb, baa, acb

(b) Explain why ρ is not an equivalence relation.

E 8.   Let σ be the transitive closure of ρ (Problem 7). Then σ is an equivalence relation (in fact, it is a congruence).

(a) Identify two distinct congruence classes of σ and list three words in each congruence class.

(b) Is every word congruent to its mirror image

(where the mirror image of a word is the word written backwards)? Explain.

P   9.   Imagine a mapping ϕ between groupoids $(A, \{\theta_1\})$ and $(B, \{\theta_2\})$ so that

$$\phi(x\theta_1 y) = \phi(y)\theta_2\phi(x)$$

Explain the relationship between this kind of mapping and congruences and admissible partitions.

E 10.   (a) Show that for the groupoid of nonnegative integers under addition the function $f: x \to$ (remainder of x divided by 3) is a homomorphism.
(b) For the homomorphism of part a, what is the admissible partition?
(c) For the homomorphism of part a, what is the congruence?

E 11.   For the homomorphism, "absolute value," which takes x into its absolute value, show the congruence and admissible partition for the groupoid $(C, \{x\})$ where C is the set of complex numbers: $C = \{x + iy \mid x, y \in R\}$.

E 12.   Two arithmetic forms are p-related if one can be substituted for another in any arithmetic expression and yield the same result. (Assume that all arithmetic expressions are fully parenthesized.)

Example
    $(a + b)$ is p-related to $(b + a)$.   #
    (a) Explain why the p-relation is a congruence.
    (b) Assume that arithmetic expressions use only addition and subtraction, and use only variable letters a, b, and c.

Example
    $((a + (b + c)) - (a + c))$ is an arithmetic expression. A p-related arithmetic expression is b.   #
List four different forms in the congruence class of $(b + c)$.

(c) What is the homomorphism associated with the p-relation?

E 13.   Let f be the following permutation on $\{1, 2, 3\}$.

$$f: 1 \to 2$$
$$2 \to 3$$
$$3 \to 1$$

Let $f^1 = f$ and $f^{k+1} = f^k \circ f$. Show the semigroup of permutations of the form f^i.

E 14.   Is the following groupoid, defined by the "multiplication" table, a monoid? A semigroup? Explain.

	a	b	c
a	a	b	c
b	b	a	a
c	c	c	c

E 15.   Let f be the following function on $\{1, 2, 3, 4\}$.

$$f: 1 \to 2$$
$$2 \to 3$$
$$3 \to 4$$
$$4 \to 4$$

Let $f^1 = f$ and $f^{k+1} = f^k \circ f$. Show the semigroup of functions of the form f^i. What is the zero of this semigroup?

P 16.   Let S_f be the groupoid whose elements are Boolean functions of two variables and whose groupoid operation is to form the function $f(f_1, f_2)$ of its two operands f_1 and f_2:

f	$\cdots$	f_2	$\cdots$
:		:	
:		:	
:		:	
f_1	$\cdots$	$f(f_1, f_2)$	

Since there are 16 functions of two variables, there will

be 16 such groupoids. Which of these groupoids are semigroups? Monoids? Groups?

E 17.   Let S be the semigroup, $\langle A, \theta \rangle$ with $A = \{a, b, c, d\}$ and θ defined by:

θ	a	b	c	d
a	a	b	c	d
b	b	c	a	d
c	c	a	b	d
d	d	d	d	d

(a) Show the extension of θ to subsets for the following cases:

(i) $\{a, b\}\theta\{c, d\} =$

(ii) $\{b, c\}\theta\{b, d\} =$

(iii) $\{a, b, c\}\theta\{a, b, c\} =$

(iv) $\{a, b, c\}\theta\{d\} =$

(b) Show that $\{a, b, c, d\}$ is an admissible partition.

(c) Show the homomorphism corresponding to the admissible partition or part b.

(d) Noting that in this case the semigroup, S, is commutative (i.e., $x\theta y = y\theta x$ for every x and y) and the homomorphic image is also commutative, explain why the homomorphic image of a commutative monoid is generally a commutative monoid.

(e) Is the converse true (i.e., suppose T is a homomorphic image of a monoid, S, and T is commutative; is S necessarily commutative)? Explain.

P 18.   (a) Let M be the monoid of functions from S to itself, where S is $\{1, 2, \ldots, N\}$. A set of *generators* of M consists of the functions:

R_N: $i \rightarrow i + 1 \bmod N$

I_N: $1 \rightarrow 2, 2 \rightarrow 1, i \rightarrow i, i \neq 1$ or 2

C_N: $i \rightarrow i + 1, i \neq N, N \rightarrow N$

Using C_4, I_4, and R_4, develop the following functions.

(i) $1 \rightarrow 1$	(ii) $1 \rightarrow 4$	(iii) $1 \rightarrow 2$
$2 \rightarrow 4$	$2 \rightarrow 3$	$2 \rightarrow 4$
$3 \rightarrow 3$	$3 \rightarrow 2$	$3 \rightarrow 1$
$4 \rightarrow 4$	$4 \rightarrow 2$	$4 \rightarrow 3$

(b) Show the semigroup generated by I_3 and C_3.

P 19. Let $M_1 = (S_1, \theta_1)$ and $M_2 = (S_2, \theta_2)$ be monoids with binary operations θ_1 and θ_2, respectively, and let $f: S_1 \rightarrow S_2$ be such that the following diagram commutes.

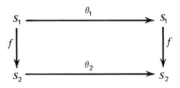

Then f is a homomorphism.
Let $M_1 = (\{a, b\}^+, \circ)$, where $\circ$ means concatenation $M_2 = (N, +)$.

(a) Let $f: a \mapsto 1$

$\qquad b \mapsto 2$

What is $f(babba)$?

(b) What is $f^{-1}(4)$?

(c) If M_1 is as just given, M_2 is any finite monoid, and a homomorphism exists from S_1 to S_2, there is a finite state machine that computes the homomorphism. Let $M_2 = (\{0, 1\}, +_2)$, where $+_2$ is a mod 2 addition and

$$f: a \mapsto 1$$
$$b \mapsto 0$$

Show a finite state machine that computes f.

E 20. $N = \{0, 1, 2, \ldots\}$. Define

$$\theta: N \times N \rightarrow N$$
$$: (x, y) \mapsto \text{lcm}(x, y)$$

(a) Show that (N, θ) is a semigroup.
(b) What is the identity, if any exists?

P 21.   Let π_1 and π_2 be partitions of S.

(a) Define $\pi_1\theta\pi_2$ to be the finest partition of which π_1 and π_2 are both refinements. What is $\{13; 24; 5; 6\} \theta \{143; 25; 6\}$?

(b) Define $\pi_1\psi\pi_2$ to the coarsest partition that is a refinement of π_1 and π_2. What is $\{132; 45; 6\} \psi \{12; 345; 6\}$?

(c) Let π be the set of partitions of S. Show that (π, θ) is a monoid.

E 22.   The permutations form a group and, hence, are solvable. Solve the following equations for the group described in Problems 3 and 4.

$$A \cdot X = B$$
$$X \cdot A = B$$
$$B \cdot X = A$$
$$X \cdot B = A$$

P 23.   A permutation is a $1-1$ onto function mapping a finite set onto itself. Show that the set of all permutations of a finite set forms a *group* under the operation of composition.

E 24.   Exhibit the group multiplication table of the group of all permutations over three elements. Is this group Abelian?

P 25.   Suppose θ is the binary operation of a finite group G, and we wish to construct a group H with the same carrier as G but with an operation ψ, where we define $a\psi b = b\theta a$, but do not change the inverse or identity functions. Is H a group? Justify your answer.

E 26.   (a) Let S_3 be the group of all permutations of three elements, $\{a, b, c\}$, under composition. Show the operation table of this group.

(b) Defining S_3 similarly for two elements, $\{a, b\}$, show the operation table for S_2.

E 27.   Let A_3 be the group of permutations, π, such that π can be formed by composition of an even number of simple

exchanges. (A simple exchange permutes just two elements and leaves the rest fixed.) Show the operation
table of A_3.

E 28. Define the following mapping from S_3 to A_3 (as defined in
Problems 26 and 27).

$$f: x \mapsto x \quad \text{if} \quad x \in A_3$$

$$x \mapsto x \circ \begin{pmatrix} 1 & 2 & 3 \\ 2 & 1 & 3 \end{pmatrix} \quad \text{if} \quad x \notin A_3$$

Is f a homomorphism? Justify your answer.

E 29. Define the following mapping from S_3 to S_2 (using the
notation of Problems 26 and 27).

$$x \mapsto \begin{pmatrix} 1 & 2 \\ 1 & 2 \end{pmatrix} \quad \text{if} \quad x \in A_3$$

$$x \mapsto \begin{pmatrix} 1 & 2 \\ 2 & 1 \end{pmatrix} \quad \text{if} \quad x \notin A_3$$

(a) Show that f is a homomorphism.
(b) What are the congruence classes?
(c) What is the admissible partition?

* 30. (a) Let P and P' be flowcharts. What should it mean to
say that P' is a homomorphic image of P?
(b) Illustrate your definition of part a on the following
pair of programs.

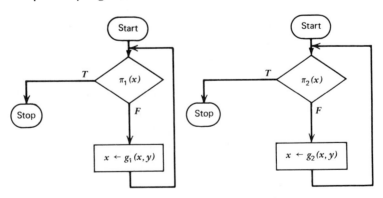

You may assume $(X, Y) \in N \times N$ and $g_1 \in N^{N \times N}$.

SUGGESTED PROGRAMMING EXERCISES

At the end of Section H, items 1, 2, and 3 are programming problems. So are the following exercises.

1. Write a program to check if a groupoid is associative.

2. Write a program that will take as input the transition function of a finite state machine and produce as output the monoid of congruence classes of the machine.

* 3. If each congruence class is considered a state, with λ being the initial state, the congruence classes may be seen as the states of a finite state machine with the same input alphabet as the original machine. In this new machine a state is final if and only if the strings in its congruence class are accepted by the original machine from which the new machine was derived. This new machine is called the semigroup machine. Write a program to check that any given machine, M, and its semigroup machine, $S(M)$, are equivalent.

formal systems

SUMMARY

We now describe the relationship among programs, machines, and computations.

A program is a specification, in a precisely defined language, of something to be computed. The specification of a programming language will be composed of several stages. The first stage is syntactic that describes what symbol sequences are to be considered programs and what internal structure is to be assigned to those programs. The next stage is semantic and specifies what the computational meaning of a program is. Finally, the pragmatic stage is concerned with the computation sequence performed by the processor that executes the program.

Initially, we discuss the syntax of several different kinds of formal language; a formal language is one in which the syntax of sentences can be developed independent of meaning. Next, a formal language for the specification of computable functions, the general recursive functions, is defined both syntactically and semantically. Since these recursive functions are based on the properties of the natural numbers and inductive definition, their description is preceded by a discussion of the properties of natural numbers and induction arguments.

We next consider the relationship of this specific computational language to the more general notion of any algorithmic language. In particular, an example of translation between two algorithmic languages is shown both as an indication of the method of proof of Church's assertion of the equivalence of all general algorithmic languages, and because the particular construction used is similar to the syntax directed translations used in compilers.

The description of Turing machines is developed next as an illustration of an extremely simple machine that can compute, and can allow proofs of unsolvability. These are particularly important, because they illustrate the limits of the algorithmic method, which is important for understanding programming methodology.

Finally, formal deductive systems are discussed with the syntax and semantics of propositional and first order-predicate logic. By now you can appreciate why the discussion of the syntax of such formal systems was preceded by the general discussion of formal syntax, and why the semantic description of

*formal systems was preceded by the discussion on the relation-
ship of algebraic systems.*

A. SYNTAX AND SEMANTICS

A grammar is ideally an algorithmic procedure for specifying a
set of sequences, any structural description of these sequences,
and their related meanings. A grammar for Algol will specify
both the set of sequences to be considered as Algol programs
and the meaning of those programs—the partial functions
computed by the program. The grammar will also provide a
description of the structure of a program.

The particular style of grammar chosen to describe a
language will influence its processing by machines and by people. Grammars for natural languages should capture the modes
of thought of the natural language and provide an accurate
description of the meanings of sentences in the language.
Grammars for computer languages should facilitate the construction of programs that compute the functions intended by
the programmer and should be processable by the computer in
a reasonable time period.

One method used to describe languages is to decompose
the description into several levels that can be processed (or
analyzed) independently. Thus, for example, a grammar for
English will specify which sentences are grammatical, such as

"Colorless green ideas sleep furiously,"

which are also meaningful, and the relationship between
different forms of the same sentence.

"John threw the baseball,"

and

"The baseball was thrown by John."

We classify the concern with *structure* as *syntax*, the description
of *meaning* as *semantics*, and the *processor* aspect as *prag-
matics*.

Definition

*The syntax of a language is the set of rules for specifying
grammatical sentences in the language.*

Definition
    The semantics *of a language is the specification of the*
meaning of sentences in the language.

Definition
    The pragmatics *of a language is the practical, psychological*
aspect of communication.

As far as possible, it is desirable to have the grammar be a
formal system in which each rule is unambiguous.

Example
    The following is a sample grammar.

S → NP – VP	VP → V
NP → Det – N	VP → V – Adv
N → boy	V → runs
N → girl	V → walks
Det → A	Adv → quickly
Det → The	Adv → slowly

This grammar is capable of generating 32 different sentences.
Each derivation for generating a sentence starts with the distinguished category symbol, S, which denotes sentence. The
other category symbols in this minigrammar are: NP (noun
phrase), VP (verb phrase), N (noun), V (verb), Det (determiner),
and Adv (adverb). Each rule consists of a left part, which is a
category symbol (a nonterminal), and a right-part, which is a
string of nonterminal symbols or a word in the language (a
terminal). The left and right sides of a rule are separated by an
arrow. At each step in the derivation, any nonterminal may be
replaced by the right-side of any rule in which that nonterminal
is the left side. Each line of such a derivation is called a
sentential form. A derivation is complete when a sentential form
consisting of terminals only is reached.
    The following is a derivation.

S
NP – VP
Det – N – VP
The – N – VP

The – girl – VP

The – girl – V – Adv

The – girl – walks – Adv

The – girl – walks – slowly

A more insightful way of representing a grammatical derivation is in the form of a derivation tree. In the derivation tree form the given derivation is represented as:

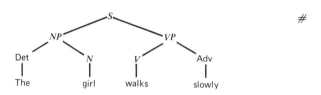

#

We can generalize on this example grammar.

Definition

A phrase structure grammar, *G, is described as G = (V_N V_T, P, S), where V_N is a finite set of nonterminal symbols, V_T is a finite set of terminal symbols, P is a finite set of* production *rules, and S is an element of V_N. Each production, p, is of the form l → r, where l and r are strings of terminal and nonterminal symbols.*

The term "phrase structure grammar" signifies that this grammar describes the structure, or phrasing, of the sentences derived in the grammar. It has been noted that such phrasing in natural language corresponds to our intuitive grouping of words, and that a speaker will tend to pause between phrases. For example, in the sentence derived above, "The girl walks slowly," the major pause will occur between the words "girl" and "walks." As can be seen from the derivation tree, this is the separation between noun phrase and verb phrase.

Definition (Derivation)

Each production of the form l → r allows us to replace l by r in any sentential form. If x = uly, y = ury, and there is a rule l → r, we write x ⊢ y (x directly derives y). The relation "⊢" is called direct derivability, *and we also write (x, y) ∈ ⊢.*

If $x \vdash x_1 \vdash x_2 \vdash \cdots \vdash x_n \vdash y$, or $x = y$, we say that x derives y, which is written $x \overset{*}{\vdash} y$. (A concise way of saying this is: "$\overset{*}{\vdash}$ is the reflexive, transitive closure of $\vdash$.")

Definition
   Let V be a finite alphabet; then V^* is the set of all finite symbol sequences over V, including the null symbol sequence. Recall that every letter in V is a sequence of length l, and that the concatenation of two sequences is a sequence. In other words, $(V^*, \{\cdot\})$ is a monoid, with the null sequence symbol λ denoting the identity of this monoid and having length 0.

Definition
   The language generated by the phrase structure grammar G, denoted $L(G)$, is

$$L(G) = \{x \mid S \overset{*}{\vdash} x \quad \text{and} \quad x \in V_T^*\}$$

The language generated is the set of sentential forms derivable from S that consists of terminal symbols only.

   It is customary to classify phrase structure grammars into four classes, depending on the restrictions on the form of the productions.

Type 0.   A grammar is of type 0 if there is no restriction on l or r.

Type 1.   A grammar is of type 1 if in each rule the length of the right side of any production is no less than the length of the left side of that production.

Type 2.   A grammar is of type 2 if the left side of each production is a nonterminal symbol.

Type 3.   A grammar is of type 3 if the left side of each production is a nonterminal symbol and the right side of each production contains at most one nonterminal symbol, in which case it is the rightmost symbol of the right side.

Each type of grammar defines a class of sets of sequences, or languages. A type *i* language is a set of sequences generated by a type *i* grammar.

The sequences generated by the grammar are referred to as the *surface structures* of the language, although it is usual in formal language studies to refer to the set of surface structures as *the* language. In addition, the grammar implicitly defines, for each surface structure, one or more *deep structures*. Ambiguous sentences in a language can be viewed as surface structures that have more than one associated, underlying, deep structure.

Example

$G = \langle V_N, V_T, P, S \rangle.$

$$V_N = \{S, T\}$$
$$V_T = \{a, b, c\}$$
$$P = \{S \to aS, \ S \to bT, \ T \to bT, \ T \to c\}$$

This grammar is of type 3, as is the language it generates, $L(G) = a*bb*c.$ #

Example

$G = \langle V_N, V_T, P, S \rangle.$

$$V_N = \{S, E, T\}$$
$$V_T = \{a, +, *, (,)\}$$
$$P = \{S \to E, \ E \to E + E, \ E \to T*T, \ T \to (E), \ T \to a\}$$

$L(G)$ is a subset of arithmetic expressions. A typical derivation tree of a *word* in $L(G)$ is:

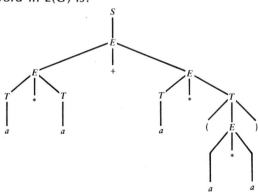

The word derived is: $a*a + a*(a*a).$ #

Note that the BNF (Backus-Naur Form) of grammar for defining a language is merely a notational variant of the phrase structure grammar of type 2. In the BNF grammar (1) each nonterminal symbol is bracketed by angular brackets allowing for names of more than one letter; (2) different rules with the same left side are combined with a single rule, the different right sides being separated by '|'; and (3) the production symbol '→' is replaced by '::='. With these changes in notation the grammar may be written:

$$\langle \text{Sentence} \rangle ::= \langle \text{Expression} \rangle$$

$$\langle \text{Expression} \rangle ::= \langle \text{Expression} \rangle + \langle \text{Expression} \rangle | \langle \text{Term} \rangle * \langle \text{Term} \rangle$$

$$\langle \text{Term} \rangle ::= (\langle \text{Expression} \rangle) | a$$

The language generated by this grammar cannot be generated by any type 3 grammar.

Informally, in a type 3 grammar any sentential form has at most one nonterminal, in which case the nonterminal is the rightmost symbol of the sentential form. In the language of arithmetic expressions, every sequence of the form $(^n a)^n$ is a valid expression. $[(^n a)^n$ is a sequence of $2n + 1$ symbols; the first n is open parentheses, the last n is close parentheses.] If the sequence $(^n a)^n$ is to be generated for arbitrarily large n, there must be an intermediate sequence of the form $(^n a)^k V$, where V must now generate exactly $n - k$ close parentheses. But because there are finitely many nonterminals in the grammar, we cannot guarantee that all, and only balanced, parentheses strings will be generated in a type 3 grammar. In a type 2 grammar this argument will not apply; the nonterminals are not restricted to be the rightmost symbols in a sentential form, so the matching parentheses pair can be generated in one production.

Example

$G = \langle V_N, V_T, P, S \rangle$, $V_N = \{S, A, B, C, D\}$, and $V_T = \{a, b, \# \}$.

$P = \{$rule 1	$S \rightarrow ACS$	rule 7	$DB \rightarrow BD$
rule 2	$S \rightarrow BDS$	rule 8	$D\# \rightarrow \#b$
rule 3	$S \rightarrow \#$	rule 9	$C\# \rightarrow \#a$
rule 4	$CA \rightarrow AC$	rule 10	$A \rightarrow a$
rule 5	$CB \rightarrow BC$	rule 11	$B \rightarrow b\}$
rule 6	$DA \rightarrow AD$		

The language generated by this type 1 grammar, $L(G)$, is

$$L(G) = \{x \,\#\, x \,|\, x \in \{a, b\}^*\}$$

$L(G)$ consists of the set of words of the form $x \,\#\, x$ where x is any word (including possibly the null word) over the alphabet $\{a, b\}$. The first three rules allow the derivation of an intermediate form in which the word x is written as $x' \,\#$, where x' is obtained from x by replacing a by AC and b by BD. The next four rules allow C's and D's to move to the right. As soon as a C or D moves adjacent to $\#$, it is transformed to an a or b. Finally, the A's and B's are also replaced by the corresponding terminals.

The following derivation shows how $ab \,\#\, ab$ is generated, using the preceding rules.

	S
rule 1	ACS
rule 2	$ACBDS$
rule 3	$ACBD\#$
rule 8	$ACB\#b$
rule 5	$ABC\#b$
rule 9	$AB\#ab$
rule 10	$aB\#ab$
rule 11	$ab\#ab$

There is no restriction in this grammar on when a rule may be applied. Thus, at the third step, rule 11 could have been used to obtain $ACBDS \vdash ACbDS$. If this had been done, it would subsequently have been impossible to eliminate the nonterminal symbol C. $\#$

A derivation in a type 1 or type 0 grammar generally cannot be described by a derivation tree, since the productions do not rewrite a single symbol. In phrase structures, grammars in which the more general rewriting productions are allowed, the derivation can be represented as a planar, directed graph. (A planar graph is a graph that can be drawn in the plane without crossovers.) In such a graph, the set of vertices with no outgoing edges is called the frontier, and the sequence of their labels of the frontier vertices from left to right, called the yield, is the sentential form derived by the graph. We define derivation graphs recursively:

1.   The graph consisting of single node ν labeled S is a derivation graph with frontier $\{\nu\}$ yield S.

2. If π_k is a production $\alpha \to \beta$, where $\alpha = \alpha_1 \ldots \alpha_e$ and $\beta = \beta_1 \ldots \beta_m$, the graph of π_k is:

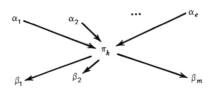

If G is a derivation graph with yield $u\alpha_1 \ldots \alpha_e v$, then G^1 is a derivation graph with yield $u\beta_1 \ldots \beta_m v$, where G^1 is obtained from G by adjoining the graph of π_k to G.

Example

The derivation graph corresponding to the given derivation of $ab \# ab$ is:

#

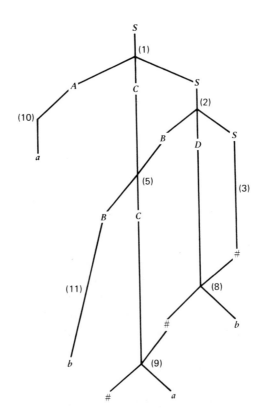

The language generated by this grammar cannot be generated by any type 2 grammar. A proof of this is beyond the scope of this text. However, this language demonstrates the inability of context-free grammars to copy symbol strings of arbitrary length in the generating process. This limitation does not allow context-free grammars to realize fully the syntax of languages such as Algol 60, in which names can be of arbitrary length and must be declared before use. Thus, if x is a variable name of arbitrary length, a program in which x occurs would include at least two occurrences of x, once where x is declared and once where x is used. This program text is a symbol string of the form $\alpha x \beta x \delta$. Context-free grammars cannot generate copies of arbitrary symbol strings in this form.

The study of formal languages allows certain questions to be precisely formulated and amenable to mathematical analysis; an informal approach would not allow such study. For example, the acceptability of a proposed sentence in English is not always a matter about which all native speakers would agree, and will often vary from one dialect to another. Among the questions that can be asked about a formal language are the following *decision problems*. (A decision problem is a problem having a yes/no answer.)

1. *Membership.* Given L and x, is $x \in L$?

2. *Finiteness.* Given L, is L finite?

3. *Regularity.* Given L, is L of type 3?

4. *Emptiness.* Given L, is L empty?

5. *Equivalence.* Given L_1 and L_2, does $L_1 = L_2$?

Table 4-1 summarizes the status of these decision problems for the classes of languages in the Chomsky hierarchy. In each case, it is assumed that the language about which the question is being asked is specified as a grammar of the same type.

Is the decision problem solvable by an algorithm?

The decision problem for a language class is decidable if there is an algorithm to determine (given a grammar in the

TABLE 4-1   DECIDABILITY OF PROBLEMS FOR CLASSES OF LANGUAGE OF THE CHOMSKY HIERARCHY.

	Type 3	Type 2	Type 1	Type 0
Membership	Yes	Yes	Yes	No
Emptiness, finiteness	Yes	Yes	No	No
Equivalence, regularity	Yes	No	No	No

appropriate class) the answer to that decision problem for that language. Let us consider the emptiness question for type 2 languages, for which an algorithm exists. The grammar, $G = (V_N, V_T, S, P)$, will generate terminal strings if and only if there are derivations starting at S that yield terminal strings. We can determine this by defining a sequence of subsets of V_N as follows.

$$V_N^0 = \phi$$

$$V_N^i = \{V \mid V \to \alpha, \alpha \text{ a string over } V_T \cup V_N^{i-1}\} \cup V_N^{i-1}$$

V_N^1 is the subset of nonterminals that can immediately be rewritten, by some productions, as terminal strings. V_N^{i+1} is the union of V_N^i and the subset of nonterminals that can immediately be rewritten, by some production, as strings of terminals or nonterminals in V_N^i. (Intuitively, if a nonterminal is a member of V_N^i, there is a derivation tree of a word in the language whose depth is i or less.) The sequence of V_N^i is finite, since $V_N^i \subseteq V_N^{i+1}$ and, if $V_N^k = V_N^{k+1}$, then $V_N^{k+l} = V_N^k$ for all l, because we never add any nonterminals to V_N^k, and V_N is itself finite. Let $\hat{V}_N = \min_i V_N^i$ such that $V_N^i = V_N^{i+1}$. Then L is nonempty if and only if S is a member of $\hat{V}_N$.

Example

Applying the emptiness algorithm to the grammar of arithmetic expressions, given earlier, with productions $\{S \to E, E \to E + E, E \to T \times T, T \to (E), T \to a\}$, we have:

$$V_N^0 = \{T\}$$
$$V_N^1 = \{E, T\}$$
$$V_N^2 = \{E, T, S\}$$
$$V_N^3 = V_N^2 = \hat{V}_N \qquad\qquad \#$$

The undecidable questions shown in Table 4-1 are all shown to be undecidable by a method of proof known as "reducibility to the halting problem for Turing machines." We will see in Section D that the halting problem is unsolvable. By very clever encoding one can develop a grammar to simulate the behavior of a Turing machine, generating a word only when the Turing machine halts. Again, the details are omitted as being beyond our scope.

It can be seen from Table 4-1 that decision problems become more difficult as we proceed from type 3 to type 0. This is a consequence of the increased descriptive power of the grammars as we proceed from type 3 to type 0. From our definitions it is clear that every language of type i is also a language of type j for each $j < i$, and we have also shown a language that is of type 1 but not of type 2, and a language that is of type 2 but not of type 3. It can also be shown that there are languages of type 0 that are not of type 1. From the fact that each language of type i is also of type j for each $j < i$, we can conclude that

1. Any decision problem decidable for the class of type i languages is decidable for the class of type j languages, $j > i$.

2. Any decision problem undecidable for the class of type i languages is undecidable for the class of type j, $j < i$.

New languages can also be defined by operations on given languages. For example, the class of expressions that are syntactically valid in both Algol and PL/I is defined as the intersection of the set of valid Algol expressions and the set of valid PL/I expressions.

If θ is an n-ary operation on the class of languages, $\theta(L_1, L_2, \ldots L_n)$ is a language. If, whenever $L_1, L_2, \ldots, L_n$ are all of type i, the resulting language, $\theta(L_1, L_2, \ldots, L_n)$, is also of type i, the class of languages of type i is said to be *closed under* θ.

Example

If L_1 and L_2 are context-free languages, they can be generated by grammars $G_1 = (V_{N_1}, V_{T_1}, P_1, S_1)$ and $G_2 = (V_{N_2}, V_{2}, P_2, S_2)$. Define a new grammar $G'_2 = (V'_{N_2}, V_{T_2}, P'_2, S'_2)$, where any nonterminals of G_2 that occur in G_1 are replaced by new symbols. It is easy to see that $L(G_2') = L(G_2)$, since the particular choice of syntactic category names is arbitrary. A context-free grammar for $L_1 \cup L_2$ is $G = (V_{N_1} \cup V_{N_2}' \cup \{S\}, V_{T_1} \cup V_{T_2}, P_1 \cup P_2' \cup \{S \to S_1, S \to S_2'\}, S)$. G includes all the nonterminals of G_1 and G_2' and an additional symbol S, which occurs neither in V_{N_1} nor in V_{N_2}'. The additional symbol, S, is the start symbol of the new grammar, G, and one of the two productions $S \to S_1$ and $S \to S_2'$ must be used to start derivations. If $S \to S_1$ is used for the initial step of a derivation, the rest of the derivation is essentially a derivation in G_1. Similarly, if $S \to S_2'$ is initially used, the derivation is essentially in G_2'. Disjoint grammars are constructed in order to assure that each derivation in G corresponds to either a derivation in G_1 or a derivation in G_2'.   #

The closure properties of the Chomsky hierarchy of language classes under union, intersection, and complement (with respect to V_T), are shown in Table 4-2.

Proposition. Let L be a class of languages closed under union and complement; L is then closed under intersection.

Proof. $L_1 \cap L_2$ can be written as $\sim(\sim L_1 \cup \sim L_2)$. Since the language class L is closed under complement, $\sim L_1$, $\sim L_2$, $\sim L_1 \cup \sim L_2$, and $\sim(\sim L_1 \cup \sim L_2)$ are all seen to be of type L.   #

Languages can also be studied as algebras, as shown in Chapter 3.

TABLE 4.2   CLOSURE PROPERTIES OF CLASSES OF LANGUAGES IN THE CHOMSKY HIERARCHY.

	Type 3	Type 2	Type 1	Type 0
Union	Closed	Closed	Closed	Closed
Intersection	Closed	Not closed	Closed	Closed
Complement	Closed	Not closed	Unknown	Not closed

Definition (The class of prefix languages)

Let T be a finite set of ranked symbols (i.e., a finite set of symbols, T, and a ranking function, $r: T \to N$). The language L is defined as follows.

(a) $r(x) = 0 \Rightarrow x \in L$; rank 0 symbols are in L.

(b) $r(x) = k$ and $w_1, w_2, \ldots, w_k \in L \Rightarrow xw_1w_2 \ldots w_k \in L$.

(c) Nothing is in L unless it follows from 1 and 2.

Example
   $T = \{a, b, c, d\}$.

$$r: a \mapsto 2, \ b \mapsto 1, \ c \mapsto 0, \ d \mapsto 0$$

Then the following words are in L:

$$\{c, d, bc, bd, bbc, bbd, acc, acd, adc, add, badbc, abcbd, \ldots\}$$

Of course, L can be defined algebraically as follows: c and d are constants; a and b are operations. If we call the carrier C, then $a: C \times C \to C$, and $b: C \to C$ by the obvious rules.

$$a(w_1, w_2) = aw_1w_2$$
$$b(w_1) = bw_1$$

An algebra defined in this way is called a *word algebra*.   #
   It is not difficult to show the following.
Proposition.   Every prefix language is a type 2 language.

Example
   $G = (V_N, V_T, P, S)$.

$$V_N = \{S\}$$
$$V_T = \{a, b, c, d\}$$
$$P = \{S \to aSS, \ S \to bS, \ S \to c, \ S \to d\}$$

$L(G)$ is the prefix language defined in the previous example.   #
   We close this section with a last example of a quite different style of grammar. These classes of grammars, known as L systems, have been studied as models of the development of biological systems.

   The interpretation of a word in an L system is that it represents the state of a filamentary organism in which each letter corresponds to the state of a cell. A rewriting rule is thus a model of development in which the cell subdivides. In the simplest L systems, the development of each cell is independent of its neighbors. Moreover, in the model all cells are evolving concurrently, so that rewriting rules are simultaneously applied to all letters of the words.[1]

---

[1]For further motivation and discussion of different L systems, see G. T. Herman and G. Rozenberg, *Developmental Systems and Languages*, Amsterdam: North-Holland Publishing Company, 1974.

Definition

An OL system *is defined as* (V_T, A, P), *where* V_T *is a finite set of (terminal symbols), A is an axiom,* $A \in (V_T)^*$, *and P is a finite set of rewriting rules of the form* $l \rightarrow r$, *where* $l \in V_T$ *and* $r \in (V_T)^*$.

At each step of the derivation each symbol in the current sentential form is replaced by the right side of a rule in which that symbol is the left side.

Definition

Let $L = (V_T, A, P)$ *be an OL system. The* sequence of sentential forms *generated by* L *is* $(S_0, S_1, \ldots)$, *where* $S_0 = A$ *and* S_i *is obtained from* S_{i-1} *by replacing each symbol in* S_{i-1} *by the right side of a rule for that symbol.*

Example

$L = (\{a, b\}, ab, \{a \rightarrow ab, b \rightarrow abb\})$.

The following sequence is generated.

ab

ababb

ababbababbabb

. . .

. . .

. . .   #

Example

$L = (\{a, b\}, a, \{a \rightarrow b, b \rightarrow ab, b \rightarrow ba\})$.

One sequence generated by *L* is:

a

b

ab

bba

abbab

. . .

. . .

. . .   #

Similar to the Chomsky hierarchy, a hierarchy of *L* systems, as well as decision problems and closure properties, has been studied. The properties of L systems are quite different from those of the Chomsky hierarchy. Indeed, each new formal language has to invent and develop its own mathematical concepts and machinery.

B. RECURSIVE DEFINITIONS AND RECURSIVE FUNCTIONS

A basic concept, central to computer science, is that computation on symbols is inherently no different from computation on numbers. Thus every *definable* set of strings (e.g., definable by a grammatical system) corresponds to a set of numbers specified by some rule of computation on integers, or natural numbers. We begin with a specification of the properties of the natural numbers as contained in the Peano axioms.

1. Zero is a natural number.
2. Each natural number has a unique successor that is a natural number.
3. If X is a subset of the natural numbers, X includes the first natural number, and the successor of any element of X is in X, then X is the set of natural numbers.
4. Every natural number is the successor of at most one natural number.
5. Zero is not a successor of any natural number.

It is easy to see that these properties apply to the natural numbers. In fact, all the properties are necessary to ensure that the system specified is (isomorphic to) the natural numbers.

Example

Consider a system in which only the first four axioms apply. In that case, any finite set of n symbols, $\{a_0, a_1, a_2, \ldots, a_{n-1}\}$ satisfies the axioms if a_0 is taken as the first number and the successor of a_i is a_p, where $p \equiv i + 1 \pmod{n}$.   #

Example

Consider a system in which only axioms 1 to 3 and axiom 5 apply. Let $\{a_0, a_1\}$ be defined with a_0 being the first natural number; let the successor of a_0 be a_1 and the successor of a_1 be a_1. It is easily verified that this system satisfies every axiom other than axiom 4.   #

Example

Axioms 1 to 4 are satisfied by the integers and by the natural numbers.   #

Exercise.   Describe a mathematical model other than the natural numbers that satisfies all axioms except possibly axiom 2. Do the same for all axioms except axiom 3.

In order to establish properties of the system of natural numbers, a sixth axiom, *the principle of mathematical induction,* is added.

Let *P* be any property of numbers. If the following two statements are true, then *P* is true of every natural number.
1.   *P* is true of 0. This is called the *basis.*
2.   If *P* is true of *n*, then *P* is true of *n*'s successor. This is called the *induction.*

Thus, in an induction proof, two things are to be shown: (1) basis—that *P* is true for 0; and (2) induction—given that *P* is true for *n* it follows that *P* is true for *n* + 1.

Example

To show that $\sum_{i=0}^{n} i = \dfrac{n(n+1)}{2}$.

$$basis \qquad \sum_{i=0}^{0} i = \frac{0.1}{2} = 0$$

$$induction \qquad \text{Assume}$$

$$\sum_{i=0}^{n} i = \frac{n(n+1)}{2}$$

to show that $\sum_{i=0}^{n+1} i = \dfrac{(n+1)(n+2)}{2}$.

$$\sum_{i=0}^{n+1} i = \sum_{i=0}^{n} i + n + 1 \qquad \text{by definition of } \Sigma$$

$$= \frac{n(n+1)}{2} + n + 1 \qquad \text{by induction assumption}$$

$$= \frac{n(n+1)}{2} + \frac{2(n+1)}{2} = \frac{(n+1)(n+2)}{2} \qquad \#$$

Example

In the Fibonacci recurrence equation,

$$x_{n+2} = x_{n+1} + x_n \qquad \text{with} \qquad x_0 = x_1 = 1$$

to show that $x_{n+2} = \sum\limits_{i=0}^{n} x_i + 1$.

$$basis \qquad x_2 = \sum\limits_{i=0}^{0} x_i + 1 = x_0 + 1 = 2$$

$$induction \qquad \text{Assume} \qquad x_{n+2} = \sum\limits_{i=0}^{n} x_i + 1$$

to show that $x_{n+3} = \sum\limits_{i=0}^{n+1} x_i + 1$

$$x_{n+3} = x_{n+2} + x_{n+1} \qquad \text{recurrence equation}$$

$$= \sum\limits_{i=0}^{n} x_i + x_{n+1} + 1 \qquad \text{by induction assumption}$$

$$= \sum\limits_{i=0}^{n+1} x_i + 1 \quad \#$$

The Peano axioms can be replaced by the following axiom.[2] The set of natural numbers, N, has zero as a member, and the successor function, $S: N \to N$. Any other set X with distinguished element, $a \in X$, and function, $f: X \to X$, has exactly one sequence, $q: N \to X$, which satisfies the equations:

$$q(0) = a, \qquad q(s(n)) = f(q(n)) \qquad \text{for all} \qquad n \in N$$

This axiom can be represented by the following diagram, which shows the recursive property of the integers.

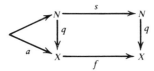

This diagram is required to commute in the sense of the homomorphism diagrams of Chapter 3.

Recursive Definition

Just as the properties of the natural numbers can be established from the Peano axioms, we should be able to define computations in terms of simpler computations. Such a system of

[2]For details see S. Maclane and G. Birkhoff, *Algebra*, Chapter II. New York: Macmillan, 1968.

definitions would constitute a "grammar" of computable functions.

Example

[Recursive definition of *addition* in terms of *successor* (successor is denoted by s).]

(a) $a + 0 = a$
(b) $a + s(b) = s(a + b)$

This definition has a *basis* and a *recursive* part. The basis defines addition of 0, and the recursive part defines addition of the successor of b in terms of addition of b.

$$4 + 3 = s(4 + 2) = s(s(4 + 1))$$
$$= s(s(s(4 + 0)))$$
$$= s(s(s(4))) = s(s(5)) = s(6) = 7 \quad \#$$

Example

(Recursive definition of *multiplication* in terms of *addition*.)

(a) $x \cdot 0 = 0$
(b) $x \cdot s(y) = x \cdot y + x$

Multiplication by 0 is directly defined, and multiplication by $s(y)$ is defined in terms of multiplication by y and addition.

$$4 \cdot 3 = 4 \cdot 2 + 4 = (4 \cdot 1 + 4) + 4 = (((4 \cdot 0 + 4) + 4) + 4)$$
$$= ((0 + 4) + 4) + 4 = \ldots = 12 \quad \#$$

The system for the recursive definition of functions generalizes these examples and provides a means of defining all known computable functions. (See Church's thesis in Section *d*.) We start with the *basis* functions.

1.   Successor function: $s(x) = x + 1$.

2.   Zero function: $0(x) = 0$.

3.   Generalized identity (or projection) function:

$$I_k^n(x_1, \ldots, x_n) = x_i \qquad k \leqq n \qquad \text{with} \qquad I_0^0 = 0$$

Example

$$I_3^6(8, 7, 2, 4, 5, 6) = 2. \quad \#$$

The basis functions provide an infinite set of functions from which other functions may be built using the following.

1. *Composition of functions.* If f is an n-ary function and $g_1, g_2, \ldots, g_n$ are k-ary functions, then $f(g_1, \ldots, g_n)$ is a k-ary function, where $f(g_1, \ldots, g_n)(x_1, \ldots, x_k) = f(g_1(x_1, \ldots, x_k), \ldots, g_n(x_1, \ldots, x_k))$.

2. *Primitive recursion.* Given an n-ary function, g, and an $(n + 2)$-ary function, h, the $(n + 1)$-ary function f, defined by primitive recursion, is

$$f(0, x_n, \ldots, x_1) = g(x_n, x_{n-1}, \ldots, x_1)$$

and $f(m + 1, x_n, \ldots, x_1) = h(m, f(m, x_n, \ldots, x_1), x_n, \ldots, x_1).$

3. *Minimization.* Given an $(n + 1)$-ary function, g, an n-ary function, defined by minimization, is

$$f(x_1, \ldots, x_n) = \mu y[g(y, x_n, \ldots, x_1) = 0]$$

where $\mu y[g(y, x_n, \ldots, x_1) = 0]$ is defined as the smallest value of y among the natural numbers, such that $g(y, x_n, \ldots, x_1) = 0$.

Using the basic functions—zero, successor, and generalized identity—the schemata for constructing new functions from given ones can be represented by the following flowcharts.

Composition

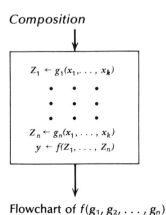

Flowchart of $f(g_1, g_2, \ldots, g_n)$

Primitive recursion:

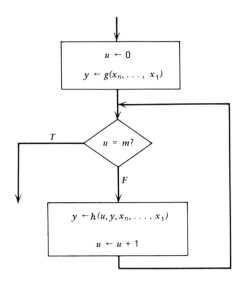

Flowchart for defining $f(0, x_n, \ldots, x_1) = g(x_n, \ldots, x_1)$
$f(m + 1, x_n, \ldots, x_1) = h(m, f(m, x_n, \ldots, x_1), x_n, \ldots, x_n)$

Minimization:

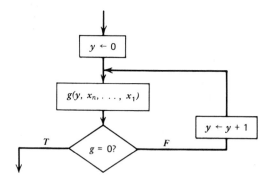

Definition
    Any function definable in terms of successor, zero, generalized identities, composition, and primitive recursion is called a primitive recursive function.

Definition

If, in addition to the primitive recursive operations, the operation of minimization is used, the function is called general recursive. If a function f is general recursive and f is total, f is called total recursive; if f is general recursive and a partial function, f is called partial recursive.

Fact.   Every primitive recursive function is a total function.
Proof.   The only possibility for a computation of a general recursive function not to terminate is for the computation to enter a minimization loop for a function that is never equal to zero. But minimization is not an allowable construct in primitive recursive definitions.   #

Example

Preceding the definition of primitive recursion and general recursion, an "informal" definition of addition was given. More formally, we have:

$$\text{plus } (0, x) = I_1^1(x)$$
$$\text{plus } (m + 1, x) = (S \circ I_2^3)(m, \text{plus } (m, x), x)$$

where $S \circ I_2^3$ is the composition of the successor and generalized identity function. Note that $S \circ I_2^3$ is a function of three variables, as required by the primitive recursive scheme.   #

Example

Multiplication, formally.

$$\text{times } (0, x) = 0(x)$$
$$\text{times } (m + 1, x) = \text{plus } (I_2^3(m, \text{times } (m, x), x), I_3^3(m, \text{times } (m, x), x))   \#$$

Example

Predecessor function.

$$\text{pred } (0) = I_0^0$$
$$\text{pred } (m + 1) = I_1^2(m, \text{pred } (m))$$

Note that I_0^0 is a constant (or function of zero variables), as required by the primitive recursive schema.

Example
Proper subtraction, defined as

$$x \stackrel{.}{-} y = \begin{array}{ll} x - y & \text{if} \quad x \geq y \\ 0 & \text{if} \quad x \leq y \end{array}$$

is primitive recursive. (See Problem 15, p. 000.) In terms of proper subtraction and addition, we can define

$$|a - b| = (a \stackrel{.}{-} b) + (b \stackrel{.}{-} a)$$

Subtraction is not primitive recursive, since it is not a total function on the natural numbers. However, it is definable as a partial function.

$$a - b = \mu y (y + b = a) \quad \#$$

Proof That Addition Is Commutative

Having defined addition in terms of successor, we can prove that this definition has properties that are familiar to us. To illustrate this, we show that addition is commutative. First, two lemmas are required. The following properties are used to define addition.

$$A: x + 0 = x$$
$$B: x + Sy = S(x + y)$$

Lemma 1

$x + 0 = 0 + x.$
Proof.   (By induction on x.)

$$\text{basis} \quad 0 + 0 = 0 + 0$$

$$
\begin{array}{lll}
\text{induction} & Sx + 0 = Sx & \text{property A} \\
& = S(x + 0) & \text{property A} \\
& = S(0 + x) & \text{induction hypothesis} \\
& = 0 + Sx & \text{property B} \quad \#
\end{array}
$$

Lemma 2

$Sy + x = S(y + x).$
Proof.   (By induction on x.)

$$
\begin{array}{lll}
\text{basis} & Sy + 0 = Sy & \text{property A} \\
& = S(y + 0) & \text{property A}
\end{array}
$$

$$
\begin{aligned}
\textit{induction} \quad Sy + Sx &= S(Sy + x) && \text{property B} \\
&= S(S(y + x)) && \text{induction hypothesis} \\
&= S(y + Sx) && \text{property B} \quad \#
\end{aligned}
$$

THEOREM. $y + x = x + y$.

Proof.   (By induction on y.)

$$
\begin{aligned}
\textit{basis} \quad x + 0 &= 0 + x && \text{lemma 1} \\
\textit{induction} \quad x + Sy &= S(x + y) && \text{property B} \\
&= S(y + x) && \text{induction hypothesis} \\
&= Sy + x && \text{lemma 2} \quad \#
\end{aligned}
$$

Proofs by Recursion

The method of proof by mathematical induction just presented has a generalization that is useful in proving properties of functions defined by recursive programs. This generalization is *the principle of structural induction on well-founded sets*. We first define well-founded set and provide some examples of well-founded sets.

The notion of a well-founded set is a generalization of the concept of a well-ordered set (Chapter 1). Recall that a well-ordered set is a linearly ordered set in which each nonempty subset has a least element under the ordering. Thus the natural numbers are a well-ordered set under usual ordering, while the integers (positive and negative) are not. Well-founding is the generalization of well-ordering, which is applicable to partially ordered sets.

Definition

A partially ordered set, P, is well-founded *if there is no infinite sequence of decreasing elements (i.e., starting from any given element of P, only a finite sequence of decreasing elements can be developed until a minimal element is reached).*

Example

Every well-ordered set is a well-founded set. $\#$

Examples

The set of ordered pairs of natural numbers is a well-founded set under any of the following orderings.

(a) $(x, y) > (x', y')$ if $x > x'$
(b) $(x, y) > (x', y')$ if $|x - y| > |x' - y'|$
(c) $(x, y) > (x', y')$ $x + y > x' + y'$ #

Examples

The set of finite sequences over an alphabet is a well-founded set under the following orders:

(a) $l_1 > l_2$ if length $(l_1) >$ length (l_2).
(b) $l_1 > l_2$ if the number of different symbols in l_1 is greater than the number of different symbols in l_2.   #

Principle of Structural Induction on Well-founded Sets. *Let S be a partially ordered set that is well-founded under the ordering relation,* $>$; *if P is any proposition and if the following two conditions hold,*

(a) *P is true of each minimal element in S (basis).*
(b) *For each element x in S, if P is true of each element in S less than x, then P is true of x (induction).*

then P is true of each element in S.

The principle of structural induction is sometimes stated only as condition (b) of the preceding principle, since condition (a) is actually redundant. That is, if x is a minimal element of S, there are no elements in S less than x, and so P is vacuously true of each element in S less than x (since there aren't any to make it false!). So we must show that P is true of x. However, in proof using structural induction, one usually follows the pattern indicated in the two condition statements of the principle.

Example

Consider the recursive program: Procedure $F(x, y)$; if $x = y$ *then* 1 *else* $F(x, y + 1)(y + 1)$, where we wish to show that $F(x, 0) = x!$. We use the ordering $(x, y) > (x', y')$ if $(x - y)$, $(x' - y') \in N$, and $(x - y) > (x' - y')$ to show, by structural induction, that when $x, y \in N$ and $x \geq y$ that $F(x, y)$ computes $x!/y!$.

basis   If $x = y$ (i.e., for minimal element), $F(x, y) = 1 = x!/y!$.
induction   If $x > y$,   then $F(x, y) = F(x, y + 1)(y + 1) =$
$\dfrac{x!}{(y + 1)!} (y + 1) = x!/y!$.

Therefore, by structural induction $F(x, y) = x!/y!$ for $x \geq y$, and $F(x, 0) = x!$.   #

Example
   Procedure $F(x, y)$; *if $x = 0$ then y else $F(x - 1, y + 3x(x - 1) + 1)$*. We use structural induction to show that for $(x, y) \in N^2$, $F(x, y + 1) = 1 + F(x, y)$.
   We use the ordering $(x, y) > (x', y')$ if $x > x'$;

basis   If $x = 0$, then $F(0, y + 1) = y + 1 = 1 + F(0, y)$.

induction   If $x > 0$, *then*

$$F(x, y + 1) = F(x - 1, y + 3x \cdot (x - 1) + 2)$$
$$F(x, y) = F(x - 1, y + 3x \cdot (x - 1) + 1)$$

and, by inductive hypothesis,

$$F(x - 1, y + 3x \cdot (x - 1) + 2) = 1 + F(x - 1, y + 3x \cdot (x - 1) + 1)$$

Thus, substituting above,

$$F(x, y + 1) = 1 + F(x, y)$$

Using this fact, it is easy to show that

$$F(x, y) = F(x - 1, y) + 3x \cdot (x - 1) + 1$$
$$= F(0, y) + \sum_{i=0}^{x-1} (3i^2 + 3i + 1)$$
$$= F(0, y) + x^3$$
$$= y + x^3 \quad \#$$

C. ALGORITHMS
   An algorithm is a computational recipe for systematic, possibly mechanical, execution. We can specify the following general characteristics of algorithms.

1.   An algorithm is given as a set of instructions of finite size.

2.   There is a computing agent, human or machine, that can react to the instructions and carry out the computations.

3.   There are facilities for marking, storing, and retrieving intermediate results of the computation.

4.   Let P be a set of instructions as in 1, and M be a computing agent as in 2. Then M interprets P in such a way that for a given input the computation is carried out in a discrete stepwise fashion without the use of continuous methods or analog devices.

5. M interprets P in such a way that a computation is carried forward deterministically without resort to random methods or devices.

The concept of an algorithm is an idealization of the underlying logic of an actual program. For example, item 5 requires that strict determinism be observed, but the requirement may contradict the possible mechanical capabilities in nature, since quantum mechanical considerations preclude complete determinism. Similarly, the interpretation of possible voltages in electronic devices as discrete is an idealization, and the internal mechanisms of the computing device may indeed rely on continuous forces.

Example (Euclidean algorithm)
This algorithm computes the greatest common divisor of two positive integers, m and n.

(1) If $m = n$, then gcd $\leftarrow m$;
else $a \leftarrow$ max (m, n)
       $b \leftarrow$ min (m, n)

(2) If b divides a, then gcd $\leftarrow b$;
else temp $\leftarrow b$
        $b \leftarrow$ remainder (a/b)
        $a \leftarrow$ temp

(3) Go to step 2.   #

Example

Carrying out the Euclidean algorithm for $m = 210$, $n = 18$ yields the following sequence of intermediate results.

$$a = 210 \qquad b = 18$$
$$a = 18 \qquad b = 12$$
$$a = 12 \qquad b = 6$$

The answer is 6.   #

An algorithm is a single list of instructions defining a computation that can be carried out on any initial data in the *domain* of the algorithm and that in each case gives the correct result. An algorithm tells how to solve not just one problem, but a whole class of similar problems.

An algorithm is often understood in a more abstract sense than that specified by the definition. Therefore, most programs for computing the greatest common divisor are described as embodiments of the Euclidean algorithm, even though they may differ in their instructions, computing agent, domains of definition, and other pragmatic aspects.

In addition to the specified requirements for an algorithm, there are a number of important pragmatic aspects: size of instruction set, complexity of computing agent, and cost of computation both in time and in storage. The organization of the algorithm for comprehensibility is also an important factor.

The methods of describing algorithms have been studied extensively. The languages that describe algorithms include:

1.   Programming languages (e.g., Algol).

2.   Flowcharts.

3.   Decision tables.

4.   Markov algorithms.

5.   Post canonical systems.

6.   Recursive functions.

7.   Lambda calculus.

8.   Turing machines.

Not all of these algorithmic languages are described in this text, but we cite the following information.

THEOREM.   Any algorithm describable by any of the preceding algorithmic languages can be algorithmically translated to any other of the algorithmic languages cited.

The general method for proving this theorem is to show how each statement in algorithmic language A can be translated to a sequence of statements in algorithmic language B.

Example
To show the translation from one algorithmic language to another, let language A be the language of structured flowcharts and language B a typical subset of an assembly programming language.

Language A is specified as follows:

(A) The following basic instructions are included:

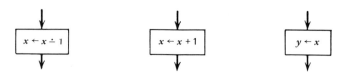

where x and y are any variable names.

(b) The following test is included:

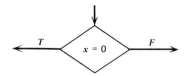

where x is any variable name.

In the following rules, let F_1 and F_2 be any struc-

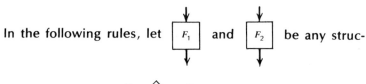

tured flowcharts and be any test.

(c) *Sequential control*:

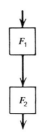

is a structured flowchart.

(d) *Conditional control*:

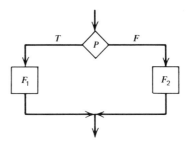

is a structured flowchart.

(e) *Iterative control*:

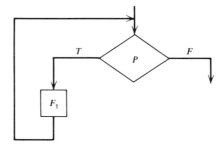

is a structured flowchart.

(f) Nothing is a structured flowchart unless it follows from rules a to e. The structured flowcharts are often represented in linear notation as follows.

Sequential control—F_1; F_2.
Conditional control—If P then F_1 else F_2 fi.
Iterative control—while P do F_1 od.

The following basic instructions are in language B.

(a) *INC X*, whose execution causes $x \leftarrow x + 1$.

(b) *DEC X*, whose execution causes $x \leftarrow x \div 1$.

(c) *MOV X, Y*, whose execution causes $y \leftarrow x$.

(d) *BR L*, whose execution causes the next step to be the one labeled *L*.

(e) *BEZ X, L*, if $X = 0$, then *BR L*; otherwise do nothing.

(f) *BNZ X, L*, if $X \neq 0$, then *BR L*; otherwise do nothing.

Each basic instruction may have a label, and the execution of the program proceeds sequentially except where a branch (*BR*) or conditional branch (*BEZ* or *BNZ*) occurs.

Let *TR(F)* denote the translation of *F*. The rules for translation can now be stated as follows.

(a) $TR(X \leftarrow X \div 1) = DEC\ X$

$TR(X \leftarrow X + 1) = INC\ X$

$TR(Y \leftarrow X) = MOV\ X,\ Y$

(b) $TR(F_1;\ F_2) = TR(F_1)$

$\qquad\qquad\qquad\ TR(F_2)$

(c) $TR(\underline{\text{If}}\ p\ \underline{\text{then}}\ F_1\ \underline{\text{else}}\ F_2\ \underline{\text{fi}}) = BNZ\ X,\ L_1$

$\qquad\qquad\qquad\qquad\qquad\qquad\ TR(F_1)$

$\qquad\qquad\qquad\qquad\qquad\qquad\ BR\ L_2$

$\qquad\qquad\qquad\qquad\qquad L1\colon TR(F_2)$

$\qquad\qquad\qquad\qquad\qquad L2\colon$

where we have assumed that *p* is "$X = 0$." The translation rule for conditional control can be seen by considering the flowchart of conditional control with labels added.

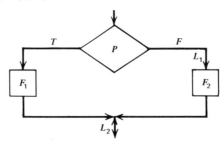

$$TR(\underline{While}\ p\ \underline{do}\ F_1\ \underline{od}) = L1:\ BNZ\ X,\ L_2$$
$$TR(F_1)$$
$$BR\ L1$$
$$L2: \qquad\qquad\qquad\qquad \#$$

Example

Using the language from the previous example, we have:

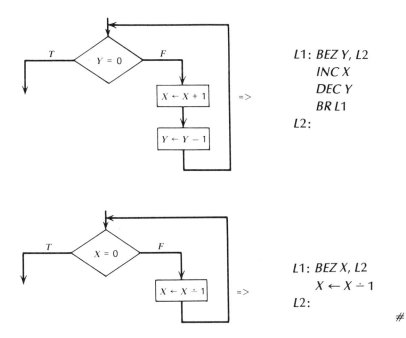

L1: *BEZ Y, L2*
   INC X
   DEC Y
   BR L1
L2:

L1: *BEZ X, L2*
   $X \leftarrow X \div 1$
L2:
   #

Exercise. Develop a structured flowchart for $A \leftarrow \max(m, n)$ and translate to language B.

The theorem stated on p. 210 says that all of the specified algorithmic languages are in some sense equivalent in their ability to describe algorithms. Although we cannot guarantee that no stronger algorithmic language is possible, none has been found, and all of the general purpose algorithmic languages known are equivalent in that an algorithm stated in one language can be translated to any other. The claim that no algorithmic language more powerful than the known languages could ever be found is known as *Church's thesis*.

CHURCH'S THESIS

Every problem having an algorithmic solution can be solved with a very simple set of instructions.

(a) Add one to A and go to the next instruction.

(b) If A is not zero, then substract one from A and go to the next instruction; otherwise go to the nth instruction;

(c) Stop and print A.

Exercise. Show how the instructions of language B in the preceding exercise can be translated to the primitive instructions in our statement of Church's thesis.

As a consequence of Church's thesis and our translation procedure, we have the equivalent statement that every problem can be solved through a flowchart (although a flowchart—even a structured one—may not be the best pragmatic solution).

Equivalent statements of Church's thesis can be formulated for each of the algorithmic languages cited.

1. Every computable function is partial recursive (i.e., every program computes a partial recursive function).

2. Every computable function is computable by a Turing machine (Section D).

D. TURING MACHINES

A Turing machine, informally, is a finite state machine with a (potentially) infinite tape. The sequencing of the machine is specified by three functions of the current state of the finite state machine and the symbol under the reading head on the tape. These three functions specify, respectively, the next state, the symbol to be written by a tape head, and the subsequent head motion (left, right, or no move). The input data are stored on the tape as follows: a unary encoding is used, so n is represented by 1^{n+1}; and a sequence of m arguments is stored as $1^{n_1+1} 0 1^{n_2+1} 0 1^{n_3+1} 0 \ldots 0 1^{n_m+1}$. The rest of the tape on either end is marked off into squares with 0's. Whenever the head moves left (or right) onto a *new* square, it contains a 0 initially. This sequence of 0's on either end is written as $\bar{0}$. Thus initial input of 3 would be recorded on the tape as $\bar{0}$ 1 1 1 1 $\bar{0}$, while an initial input sequence (2, 4, 1) would be

$$\bar{0}\ 1\ 1\ 1\ 0\ 1\ 1\ 1\ 1\ 0\ 1\ 1\ \bar{0}$$

In addition, the state is recorded directly below the symbol being scanned and indicates the position of the read head. The tape, state, and head positions are called the configuration. The starting configuration with a 3 on tape is:

$$\bar{0}\ 1\ 1\ 1\ \bar{0}$$
$$q_0$$

The machine starts computing with its head on the leftmost one and ends its computation when it first (if ever) arrives in a special state—the halt state. The result of the computation is taken to be the number of 1's left on the tape.

Example

The following Turing machine adds 1 to the number stored on tape. (In this example and others q_0 will denote the initial state of the finite state machine component of the Turing machine, q_1 will denote a final "accepting" state, and q_2 will denote a final "rejecting" state. In other formulations of Turing machines the final state is always accepting, and rejection of an input is only by an unending computation. It is easy to see that once the machine enters state q_2, it could be forced to cycle endlessly.) We specify the Turing machine by a table of "quintuples," which includes the three functions next state, symbol to be written, and head motion.

Current State	Current Symbol	Next State	Next Symbol	Head Motion
q_0	1	q_3	1	Left
q_3	0	q_1	1	No move

An initial configuration of

$$\bar{0}\ 1\ 1\ 1\ \bar{0}$$
$$q_0$$

leads to the next configuration of

$$\bar{0}\ 0\ 1\ 1\ 1\ \bar{0}$$
$$q_3$$

and, finally,

$$\bar{0}\ 1\ 1\ 1\ 1\ \bar{0}$$
$$q_1$$

Using the symbol $\underset{\text{TM}}{\vdash}$ to denote a single step of the Turing machine, and $\underset{\text{TM}}{\overset{*}{\vdash}}$ to denote 0 or more steps, we have:

$$\bar{0}\ 1\ 1\ 1\ \bar{0} \vdash \bar{0}\ 0\ 1\ 1\ 1\ \bar{0} \vdash \bar{0}\ 1\ 1\ 1\ 1\ \bar{0}$$
$$q_0 \qquad\qquad q_3 \qquad\qquad q_1$$

and

$$\bar{0}\ 1\ 1\ 1\ \bar{0} \overset{*}{\vdash} \bar{0}\ 1\ 1\ 1\ 1\ \bar{0}$$

$$q_0 \qquad\qquad q_1 \quad \#$$

It is clear that Turing machines compute partial functions since, for a given input configuration, the next configuration is uniquely determined. What is less obvious but also true is that all computable functions are computable by Turing machines.

The detailed construction of a Turing machine is an exercise in programming in a very low-level machine language. To facilitate this programming, it is useful to start with a flowchart representation of the solution using the structured flowchart components. In translating from the flowchart to the Turing machine quintuples, states of the Turing machine will correspond to labels of the flowchart. The start is labeled q_0, each accepting node is labeled q_1, and each rejecting node is labeled q_2. Each test condition is given a new label.

Example

Program a Turing machine to test if the input is of the form $2^q - 1$.

The input will be of the form $2^q - 1$ if there are 2^q 1's on the tape. Informally, the head will first move left and record a "#" at the left end of the input; it will then move to the right end of the input and record a "#." After this initialization, the head

will make repeated passes over the tape, erasing alternate "1's."
The necessary and sufficient condition that the number of 1's on
the tape be initially a power of 2 is that the head see an odd
number of 1's only when there is exactly one "1" on the tape.
We can sketch the solution as follows.

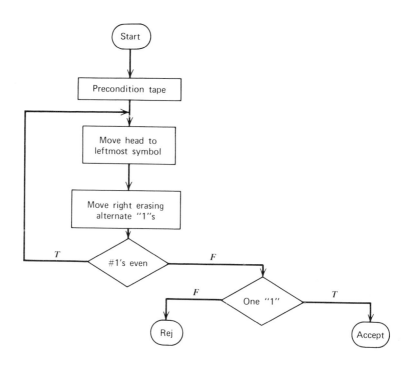

In developing the details of the solution, we expand each
component in a topdown manner.

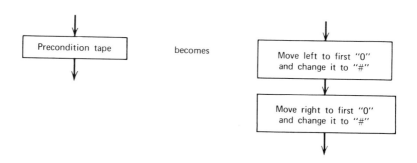

Anticipating the development of the quintuple representation of this Turing machine, we expand these components to elemental tests and actions.

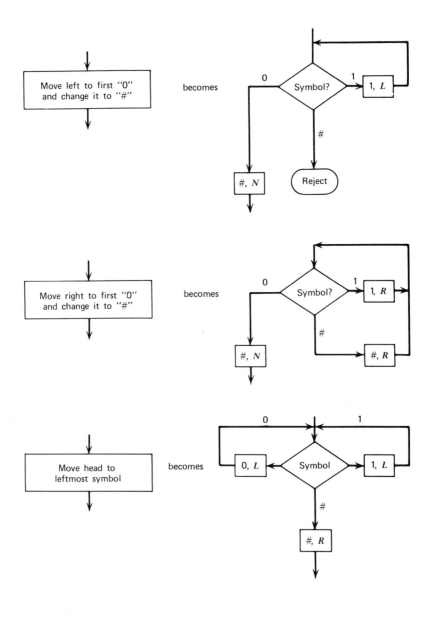

We can combine these into:

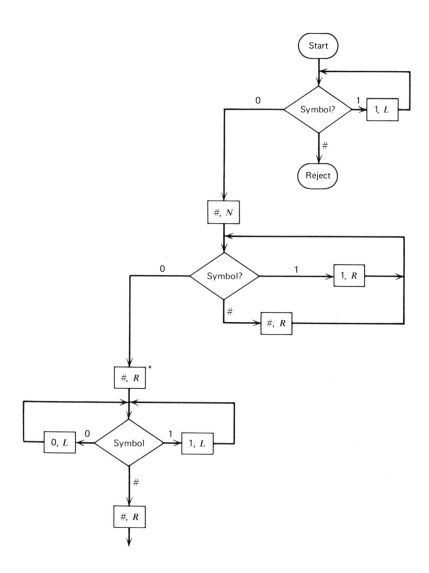

*Here we have simplified by requiring the head to move right immediately instead of checking for the second "#."

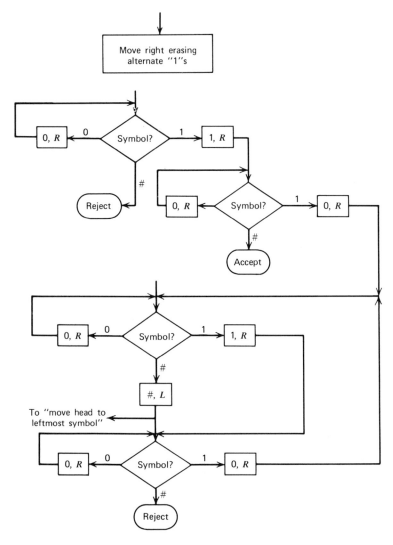

The Turing machine construction is now essentially complete. The components may now be connected into a single flowchart and the test, accept, and reject components may be labeled with states, as indicated. Then the "quintuples" may be read directly from the complete flowchart.   #

Exercise. Assemble the completed flowchart and develop the Turing machine quintuples in tabular form.

Fact. All partial recursive functions are computable by Turing machines.

Fact. Turing machines with several tapes can be simulated by single-tape Turing machines.

Fact. There exist Universal Turing machines that, given a "program" as an additional input argument, can simulate any other Turing machine.

Unsolvability of the Halting Problem for Turing Machines

The proof is by contradiction. Assume that some Turing machines, D, can decide whether any given Turing machine computation will halt. D will arrive at its conclusion by analyzing the description of the machine, T, and its tape input, t. Let us call the description of T, d_T. Then the computation of D can be visualized as:

Now D is to decide if T halts on t and, by assumption, D can decide this for any machine and any tape. Now, since D can determine if T halts on t, it is at the same time determining if T does not halt on t. Next let us modify D as follows, calling our new machine D': (1) D' will start with a description, d_T, of the machine, T, and make a copy of d_T to serve as the input tape; (2) D' will then simulate D on the description-input pair, (d_T, d_T); and (3) when the simulation of D is complete, *in a finite time*, D' will halt if D indicates that T would not halt on d_T, and D' will go into an infinite loop if D indicates that T would halt on d_T. (Remember that D was assumed to be able to analyze *any* combination of specified machine T with description d_T and tape t.)

Now we ask what D' will do when applied to D'. Suppose D' halts on its own description; then D' must loop. Suppose D' loops on its own description; then D' must halt. Hence, D' halts if and only if D' does not halt.

The assumption that leads to this contradiction is that a machine such as D exists. *Hence, we conclude that a machine such as D is a logical impossibility.*

Now, if Church's thesis that every algorithm can be computed by a Turing machine is correct, it follows that there is no algorithm to solve the halting problem.

It is possible to try to determine if a given Turing machine halts on a specified input tape. One can simulate the Turing machine in question and if, indeed, it does halt, we will find out in a finite amount of time. However, what we cannot determine in a finite amount of time is if T has an unending computation.

E. FORMAL DEDUCTIVE SYSTEMS

Formal logic is a calculus, or symbolic form of computation, for the reasoning involved in discourse. With such a calculus, extended proofs or chains of reasoning can be analyzed in an algorithmic manner to ascertain that all the assumptions are consistent with each other and that the conclusions are entailed by the assumptions.

Example

Consider the following premises.[3]
(a) Babies are illogical.
(b) Nobody is despised who can manage a crocodile.
(c) Illogical persons are despised.
Consider the following conclusions.
(a) Babies cannot manage crocodiles.
(b) Crocodile managers are not babies.   #

Example

Consider the following premises.[4]
(a) I trust every animal that belongs to me.
(b) Dogs gnaw bones.
(c) I admit no animals to my study unless they beg me to do so.
(d) All the animals in the yard are mine.
(e) I admit every animal that I trust into my study.
(f) The only animals that are really willing to beg when told to do so are dogs.

[3]From Lewis Carroll, *The Complete Works of Lewis Carroll*, New York: The Modern Library.
[4]Ibid.

Consider the following conclusion.

(a) All the animals in the yard gnaw bones.   #

The logical system discussed in this section is known as the propositional, or sentential, calculus. In this system, each propositional symbol represents a statement, and the formulas of the system represent composite sentences.

The *syntax* of the sentences, known as *well-formed formulas* of the sentential calculus, is a BNF grammar with an alphabet of arbitrary size.

1.   The propositional symbols $A, B, C, \ldots$ are well-formed formulas.

2.   If W_1 and W_2 are well-formed formulas, then so are:

   (a) $\neg W_1$ read "not W_1"
   (b) $(W_1 \& W_2)$ read "W_1 and W_2"
   (c) $(W_1 \vee W_2)$ read "W_1 or W_2"
   (d) $(W_1 \supset W_2)$ read "if W_1 then W_2"

3.   Nothing else.

Note that in item 2, parts b, c, and d, the parentheses are a part of the syntax of the well-formed formula.

Example

The sequence of symbols $A \supset B \vee C$ is not a well-formed formula, but $(A \supset (B \vee C))$ and $((A \supset B) \vee C)$ are well-formed.   #

Reformulating the syntax of well-formed formulas as a BNF grammar, we have:

$$\langle WFF \rangle ::= A|B| \ldots |\neg \langle WFF \rangle|(\langle WFF \rangle \langle \theta \rangle \langle WFF \rangle)$$

$$\langle \theta \rangle ::= \wedge | \vee | \supset$$

An alternate syntax for well-formed formulas is the *prefix* form, in which operators precede operands. Prefix form well-formed formulae are specified by:

1.   The propositional symbols, $A, B, C, \ldots$ are prefix form, well-formed formulas.

2.   If $X_1, X_2, \ldots, X_n$ are prefix form, well-formed formulae, and θ is an n-ary operator, then $\theta X_1, X_2, \ldots, X_n$ is a prefix form,

well-formed formula. ($\supset$, $\wedge$, $\vee$ are binary operators; $\neg$ is a unary operator.)

3. Nothing else is a prefix form, well-formed formula.

In BNF notation, the specification of prefix form, well-formed formulas is:

$$\langle PFF \rangle ::= A|B| \ldots | \neg \langle PFF \rangle | \langle \theta \rangle \langle PFF \rangle \langle PFF \rangle$$
$$\langle \theta \rangle ::= \wedge | \vee | \supset$$

A pair of programs, in recursive form, are readily written to convert from $\langle WFF \rangle$ to $\langle PFF \rangle$ and back. The translation from $\langle WFF \rangle$ to $\langle PFF \rangle$ is given by:

$X \leftarrow$ TRANSLATE (W):
    if $W = (W_1 \theta W_2)$ then $X \leftarrow \theta$ TRANSLATE (W_1) TRANS-
LATE (W_2);
        if $W = \neg W_1$ then $X \leftarrow \neg$ TRANSLATE (W_1);
        if $W \in \{A, B, C, \ldots\}$ then $X \leftarrow W$.

TRANSLATE is a bijection from the class of well-formed formulas to the class of prefix form well-formed formulas. Thus, an inverse function, BACKTRANSLATE, exists that can be programmed as:

$W \leftarrow$ BACKTRANSLATE (X):
    if $X = \theta X_1 \ X_2$ then $W \leftarrow$ (BACKTRANSLATE (X_1) θ
BACKTRANSLATE (X_2));
        if $X = \neg X_1$ then $W \leftarrow \neg$ BACKTRANSLATE (X_1);
        if $X \in \{A, B, C, \ldots\}$ then $W \leftarrow X$.

Informally, in each logical system there are two methods of deriving conclusions. In the proof system method, one proceeds to develop new statements from given ones by some fixed set of symbol manipulation rules. In the semantic method, one develops the meanings of statements. Ideally, one would like every conclusion derivable in the semantics to be provable, and vice versa.

The proof system of a formal logical system is a set of rules for developing the conclusions from a given set of premises by symbolic operations. Such a symbolic calculus can develop the representation of the conclusion from the premises without regard for the meaning of the symbols. The criteria that a proof system should possess are *consistency* and *completeness*. If a proof system is *consistent*, no unjustified conclusions can be drawn from the *premises*.

A proof system is *complete* if every conclusion justified by the semantics is derivable in the proof system. (A conclusion justified semantically is said to be *valid*.)

The semantics of the propositional calculus is that each formula denotes a proposition that is either *true* or *false*; these are denoted *T* and *F*, respectively. In interpreting a set of formulas, the sentential variables are assigned values. The semantics of the composite formulas are then given by the following functions.

TRUTH TABLE SPECIFICATION OF THE FUNCTIONS DENOTED BY CONNECTIVES OF THE PROPOSITIONAL CALCULUS.

A	B	$\supset AB$	$\lor AB$	$\land AB$	$\neg A$
T	T	T	T	T	F
T	F	F	T	F	F
F	T	T	T	F	T
F	F	T	F	F	T

Example

Let *A* denote *it is raining* and *B* denote *it is Tuesday*. Then *A* is true if, indeed, it is raining, and *B* is true if it is Tuesday. $\lor AB$ is true if it is raining or it is Tuesday, or both; otherwise it is false. $\land AB$ is true if it is Tuesday and it is also raining; otherwise, $\land AB$ is false. $\neg A$ is true only if it is not raining. Finally, $\supset AB$ is true except when *A* is true and *B* is false. In this case $(A \supset B)$ is taken to be true unless it is raining on a day other than Tuesday. (Although the colloquial use of the sentence, *If it is Tuesday then it is raining*, may imply a causal relationship, its usage in formal logic as a strictly truth functional form is so dominant that we will retain it.)   #

Definition
   An interpretation *of set of formulas of the propositional calculus is an assignment of truth values to the propositional variables and the subsequent assignment of truth values to the formulas according to the truth functions denoted by the connectives.*

The truth value associated with a proposition is computable from the truth assignments of the propositional variables. Assume that a function TVAL gives the truth value of the sentential variables. Call the set of sentential variables S_0. Then we can write the following recursive program.

VALUE (X):
   if $X = \theta X_1 X_2$ <u>then</u> $f(\theta)$ VALUE (X_1) VALUE (X_2);
   if $X = \neg X_1$ <u>then</u> $f(\neg)$ VALUE (X_1);
   if $X \in S_0$ <u>then</u> TVAL (X);

where $f(\theta)$ is the truth function for the binary operator θ.

Example
   $X = \neg \wedge A \vee A \neg B$, where TVAL $(A) = T$, TVAL $(B) = F$. Then VALUE $(\neg B) = T$, VALUE $(\vee A \neg B) = T$, VALUE $(\wedge A \vee A \neg B) = T$, and VALUE $(\neg \wedge A \vee A \vee B) = F$.   #
   Note that the evaluation of a logical formula is a homomorphism from the set of symbol strings to the algebra of truth values, which can be seen by inspecting the program for computing VALUE. (The translation programs for $\langle WFF \rangle$'s to $\langle PFF \rangle$'s are also homomorphisms.)
   A prefix form formula X_1 is a tautology if X_1 is true for all functions TVAL that assign truth values to the propositional symbols. A set of sentences in $\langle PFF \rangle$, $\{X_i\}$, is said to imply a formula Y, written $\{X_i\} \models Y$ if every function TVAL that makes each X_i true also makes Y true.
FACT. $\{X_i\} \models Y \Leftrightarrow (((X_1 \& X_2) \ldots \& X_n) \supset Y)$ is a tautology. Since we we can evaluate the right side using VALUE, for each possible TVAL, we have in principle a decision procedure for implication.
   First, let us convert all elements of $\langle X \rangle$ to $\langle Y \rangle$ as follows where $\langle Y \rangle$ uses only the logical symbols $\neg$, &, and $\vee$.

CONVERT (X):

    <u>if</u> $X = \& X_1 X_2$ <u>then</u> $Y = \&$ CONVERT (X_1) CONVERT (X_2)

    <u>if</u> $X = \vee\, X_1 X_2$ <u>then</u> $Y = \vee$ CONVERT (X_1) CONVERT (X_2)

    <u>if</u> $X = \supset X_1 X_2$ <u>then</u> $Y = \vee \neg$ CONVERT (X_1) CONVERT (X_2)

    <u>if</u> $X = \neg X_1$

                  <u>then</u> $Y = \neg$CONVERT (X_1)

    IF $X \in S$ <u>then</u> $Y = X$

After we have converted X to Y in $\{\neg, \vee, \wedge\}$-form, we transform Y to Z in standard form as follows.

    $Z \leftarrow$ TRANSFORM (Y):

        <u>if</u> $Y = \theta Y_1 Y_2$ <u>then</u> $Z \leftarrow \theta$ TRANSFORM (Y_1) TRANSFORM

(Y_2)

        <u>if</u> $Y \in S$ <u>then</u> $Z \leftarrow S$

        <u>if</u> $Y = \neg Y_1$ <u>then</u> DO, <u>if</u> $Y_1 = \neg Y_2$ <u>then</u> $Z \leftarrow$ TRANS-

FORM (Y_2);

                      <u>if</u> $Y_1 \in S$ <u>then</u> $Z \leftarrow \neg S$;

                      <u>if</u> $Y_1 = \wedge Y_2 Y_3$ <u>then</u> $Z \leftarrow \vee$ TRANSFORM $(\neg Y_2)$

                                      TRANSFORM $(\neg Y_3)$;

                      <u>if</u> $Y_1 = \vee Y_2 Y_3$ <u>then</u> $Z \leftarrow \wedge$ TRANSFORM $(\neg Y_2)$

                                      TRANSFORM $(\neg Y_3)$;

After TRANSFORM, the result is a string with no two successive occurrences of $\neg$. Now we define TASSERT as follows.

TASSERT (X):

    <u>if</u> $X = \vee X_1 X_2$ <u>then</u> $Y =$ TASSERT $(X_1) \cup$ TASSERT (X_2);

    <u>if</u> $X = \wedge X_1 X_2$ <u>then</u> $Y =$ TASSERT $(X_1) \cap$ TASSERT (X_2);

    <u>if</u> $X \in S$ <u>then</u> $Y = \{X\}$

    <u>if</u> $X = \neg X_1$ <u>then</u> DO;

        <u>if</u> $X_1 \in S$ <u>then</u> $Y = \{\neg X_1\}$

        <u>if</u> $X_1 = \neg X_2$ <u>then</u> $Y =$ TASSERT (X_2);

        <u>if</u> $X_1 = \& X_2 X_3$ <u>then</u> $Y =$ TASSERT $(\neg X_2) \cup$ TASSERT $(\neg X_3)$;

        <u>if</u> $X_1 = \vee X_2 X_3$ <u>then</u> $Y =$ TASSERT $(\neg X_2) \cap$ TASSERT $(\neg X_3)$;

TASSERT (X) generates the set of all assertions about the set of sentential variables, S_0, which makes X true. If TASSERT $(\neg X) = \phi$, then X is a tautology.

Example

$$X = (A \supset (B \supset A)) \qquad \text{CONVERT } (X) = \supset A \supset BA$$
$$X_1 = \text{TRANSFORM (CONVERT } (X)) = v \neg A \ v \ \neg BA$$
$$\text{TASSERT } (X_1) = \text{TASSERT } (\neg A) \cup \text{TASSERT } (\ v \ \neg BA) =$$
$$\{\neg A\} \cup (\{\neg B\} \cup \{A\}) = U \quad \#$$

Our programs like many recursive programs, have attempted to achieve a global view of the process instead of including details of sequencing (or backtracking), which would be added in an elaboration of these programs.

A Justification of Venn Diagrams

In Chapter 1, a number of formulas for manipulating expressions denoting sets were justified by reasoning about Venn diagrams. Such reasoning can be formalized, as a logical calculus. We do this for the two-variable case.

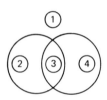

In this Venn diagram there are four disjoint regions: r_1, r_2, r_3, and r_4. Let $U \underset{\Delta}{=} r_1 + r_2 + r_3 + r_4$. The rules of the system are:

rule 1 $R(x) = r_2 + r_3$

rule 2 $R(y) = r_3 + r_4$

rule 3 $R(\neg E) = U - R(E)$ for any expression E

rule 4 $R(E_1 \cap E_2) = \Sigma r_x$

           r_x in E_1 and r_x in E_2 for any expressions E_1 and E_2

rule 5 $R(E_1 \cup E_2) = \Sigma r_x$

           r_x in E_1 or r_x in E_2

rule 6 $E_1 = E_2 \Leftrightarrow R(E_1) = R(E_2)$

Example

To show that $\neg(x \cup y) = (\neg x \cap \neg y)$:

(a) $R(x) = r_2 + r_3$ rule 1

(b) $R(y) = r_3 + r_4$ rule 2

(c) $R(x \cup y) = r_2 + r_3 + r_4$ rule 5, and (a) and (b)

(d) $R(\neg(x \cup y)) = r_1$ rule 3, and (c)

(e) $R(\overline{}x) = r_1 + r_4$ rules 1 and 3
(f) $R(\overline{}y) = r_1 + r_2$ rules 2 and 3
(g) $R(\overline{}x \cap \overline{}y) = r_1$ rules 4, and (e) and (f)
(h) $\overline{}(x \cup y) = (\overline{}x \cap \overline{}y)$ rule 6 and (d) and (h)   #

Predicate Logic

In our brief discussion of propositional logic, we have seen that this form of logic can be used to analyze arguments that involve relations among propositions, where the force of the argument depends only on the truth value of the propositions and not on their internal structure. Since not all sequences of statements in a logical presentation or mathematical argument can be reduced to statements in propositional logic, the propositional logic is extended to include variables, predicates, and quantifiers.

A complete treatment of predicate logic is beyond our scope, so we will sketch very briefly how the developments parallel and extend what we have done in the propositional logic.

First, one develops a detailed syntax of predicate logic. In prefix form, this is:

A set of variables: $\{X_i\}$

A set of function symbols: $\{f_j^n\}$, where the superscript indicates rank and $\{f_j^0\}$ are constants.

A set of predicate symbols: $\{P_k^n\}$, where the superscript indicates rank and $\{P_k^0\}$ are logical constants.

A term t is any variable, or any n-ary function symbol followed by n terms.

An atomic formula is any n-ary predicate, followed by n terms.

A formula is an atomic formula, or an n-ary logical connective followed by n formulas, or a quantifier symbol, followed by a variable symbol, followed by a formula. A quantifier symbol is $\forall$ or $\exists$.

Examples

(In no particular order.)

variables: x_1, x_2, x_3

function symbols: f_1^0, f_2^0, . . . , f_1^1, f_2^1, . . . , f_1^2, f_2^2, . . .

predicate symbols: P_1^0, P_2^0, . . . , P_1^1, P_2^1, . . . , P_1^2, P_2^2, . . .

terms: x_1, f_1^0, $f_1^1 x_1$, $f_1^2 x_1 f_1^0$, $f_2^1 f_1^2 x_1 f_1^1 f_3^0$

atomic formulas: P_1^0, P_2^0, . . . , $P_1^1 f_1^0$, $P_3^2 f_1^2 x_1 x_2 f_0^0$

formulas: $P_1^1 f_1^0$, $\exists\, x_1 P_1^1 f_1^0$, $\exists\, x_1 \forall\, x_2 P_2^2 x_2 x_1$ #

To complete this definition of syntax, we need the following.

A subformula, which is a substring of a formula that is itself a formula.

An occurrence of a variable, x, in a formula, which is bound if it occurs in a subformula whose prefix is $\exists x$ or $\forall x$. If a variable is not bound in a formula, it is free.

A sentence, which is a formula with no free variables.

In order to be able to assign truth values, we need the notion of a sentence, since formulas with free variables, such as $x = 3 + y$, are true for some assignments of values to the variables and false for other assignments of values to the variables.

An interpretation of a set of sentences in the logical system is based on a universe of elements, U, a set of ranked functions closed on those elements (i.e., an algebraic system), and a homomorphism from variable free terms in the logical system to the algebraic system. The predicates are interpreted as functions from $U^n \rightarrow \{T, F\}$. Each formula containing free variables cannot be evaluated until the variables are assigned values in U. A formula, $(\exists\ x)p$, in which x is the only free variable, is true if and only if $p(x)$ is true for some assignment of x in U. A formula $(\forall\ x)p$ in which x is the only free variable is true if and only if p is true for all assignments of x in U.

The semantic notion of validity is now that a sentence s can be validly deduced from the set of sentences S if every interpretation that makes all sentences in S true also makes s true.

A proof system for the predicate calculus should, if possible, have the desirable properties of completeness and consistency. Several such systems exist and have been mechanized. A typical proof is shown by the following example.

Example
  As in the main discussion, we do not give a full set of rules of the deduction system but illustrate one, hopefully self-explanatory, proof: $((\forall x)P(x) \supset (\exists x)(P(x) \vee Q(x)))$, where we have used the infix notation for ease of understanding.

(a) $(\forall x)P(x)$ assumption

(b) $P(a)$ instantiation: if $(\forall x)P(x)$ is true, then certainly $P(a)$ is true for any individual a

(c) $P(a) \vee Q(a)$ if $P(a)$ is true, then $P(a) \vee Q(a)$ is true

(d) $(\exists x)(P(x) \vee Q(x))$ a

(e) $((\forall x)P(x) \supset (\exists x)(P(x) \vee Q(x)))$ having assumed $(\forall x)P(x)$, we were able to prove $(\exists x)(P(x) \vee Q(x))$, so we may write $((\forall x)P(x) \supset (\exists x)(P(x) \vee Q(x)))$ #

  An additional property that would be desirable is *decidability*. A logical system is *decidable* if, given a sequence of sentences, S, there is an algorithm to determine if a given sentence, s, is a valid conclusion. In general, the predicate logic is undecidable, which means that the algorithmic method, although it can ultimately find a proof for every true proposed theorem, cannot guarantee finding an interpretation that makes all sentences in S true while making s false if S does not imply s.

REVIEW

This chapter opens with a survey of some formal linguistic notions. The syntactic structure of the languages is emphasized primarily because syntax is better understood than semantics. However, in describing the structure and meaning of expressions in a language, the syntax and semantics are two aspects of a composite picture, and the totality of information conveyed by syntax and semantics as well as the nature of this decomposition is important.

  Proceeding from these different styles of grammar, we continue with a grammar of computation—the general recursive description of the class of computable functions. We show how to translate these general recursive definitions to flowcharts. A related issue that is discussed is the use of proof by recursion as a generalization of the proof by induction method, which is introduced in Chapter 0 and expanded here.

The formal specification of computations is then related to the intuitive concept of the algorithmic method, and a further example of translation from one syntax to another is given—in this case, from flowcharts to assembly language. Church's thesis is introduced; this is the principle that any one of a number of methods of algorithmic specification can be translated to other equivalent forms, and that these forms really capture the intuitive notion of computation.

We describe Turing machines and prove the unsolvability of the halting problem for Turing machines. It is particularly important to appreciate the limitations of the algorithmic methods at the same time that one learns of the formidable efficacy of the method.

The chapter concludes with an introduction to formal logic. The propositional calculus is described as a formal system, and its syntax and semantics are given. Also, the relationship between the proof system and the semantics of the logic is discussed. Finally, a brief sketch is drawn of how these ideas can be extended to predicate calculus.

PROBLEMS

E 1. Give a BNF (or type 2) grammar for the set of all words, S, consisting of a's and b's such that the number of a's in any word is equal to the number of b's.

P 2. Is a language, specified by a BNF grammar, an algebraic system? Explain.

E 3. (a) Write a context-free grammar for the language

$$L = \{a^n b^n c^m \mid n, m \geq 1\} \cup \{a^p b^q c^q \mid p, q \geq 1\}$$

(b) Show a derivation of $a^3 b^3 c^3$ in your grammar. (Be sure to annotate both your grammar and derivation for ease of grading.)

E 4. Write a context-free grammar for

$$L = \{x \mid \text{No. of } a\text{'s in } x \geq \text{No. of } b\text{'s in } x\}$$

E 5. Let $\{p_1, p_2, \ldots, p_n\}$ be a finite set of predicates and let $\{f_1, \ldots, f_m\}$ be a finite set of functions. A structured program is defined as follows:

- Each of f_i is a structured program.
- If x and y are structured programs, $(x; y)$ is a structured program.

If x and y are structured programs and z is a predicate,

$$\text{IF } z \text{ THEN } x \text{ ELSE } y$$

is a structured program.

- Nothing is a structured program unless it is so by A and B.

(a) Give a type 2 grammar for the set of structured programs over $\{p_1, p_2\}$ and $\{f_1, f_2\}$.

(b) Show the derivation of

$$\text{WHILE } p_1 \text{ DO IF } p_2 \text{ THEN } ((f_1; f_2); f_1) \text{ ELSE } (f_2; (f_1; f_2))$$

in your grammar of part a.

E 6. Write BNF grammars for the following.

(a) Set of palindromes over $\{a, b\}$.

(b) Set of balanced parenthesis strings.

(c) Set of binary numbers divisible by 3.

E 7. For the following grammar, G, which of these strings are not accepted:

(a) *abccd* (b) *dbeacba* (c) *babcdab* (d) *abceccd*?

$$G: S \rightarrow aST$$

$$\begin{array}{lll} S \rightarrow bS & T \rightarrow Td & T \rightarrow c \\ S \rightarrow c & T \rightarrow eST & \end{array}$$

E 8. Let $f(0) = 1$, $f(1) = 1$, and $f(n + 1) = f(n) + f(n - 1)$. Using mathematical induction, show that

$$f(2n + 1) = \sum_{i=0}^{n} f(2i)$$

$$f(2n + 2) = \sum_{i=0}^{n} f(2i + 1) + 1$$

E 9. Show by mathematical induction that:

(a) $\displaystyle\sum_{k=0}^{n} (2k + 1) = (n + 1)^2$

(b) $\displaystyle\sum_{i=0}^{n} S^i = \frac{S^{n+1} - S}{S - 1}$ $S \neq 0$

E 10. Prove by mathematical induction:

(a) $\displaystyle\sum_{i=1}^{n} i^2 = \frac{n \cdot (n + 1)(2n + 1)}{6}$

(b) $\displaystyle\sum_{i=1}^{n} i^3 = \left[\frac{n \cdot (n + 1)}{2}\right]^2$

P 11. Give a primitive recursive definition of $f(n)$ defined by $f(0) = 1$, $f(1) = 2$, and

$$f(n + 1) = f(n) + f(n - 1)   n \geq 1$$

Be sure to use only primitive recursive operations.

P 12. Give a recursive definition of $\lfloor \log_2 n \rfloor$. (If you can, do this primitive recursively. It is possible.)

P 13. Is the function "square root": $N \to N$, defined by "square root" $(n) \to \lfloor \sqrt{n} \rfloor$ primitive recursive? Justify your answer.

E 14. Give a primitive recursive definition of the following function.

$$f(0) = 7$$
$$f(1) = 23$$
$$f(2) = 11$$
$$f(3) = 19$$
$$f(n) = n   n \geq 4$$

E 15. The natural numbers are not closed under subtraction. Explain. Define a form of subtraction $a \dot- b$ as a primitive recursive function, where

$$a \dot- b = \begin{cases} a - b & \text{if } a \geq b \\ 0 & \text{if } a < b \end{cases}$$

E 16. A predicate is a function, $f: D \to \{\text{true}, \text{false}\}$. Show that the predicate *is-even* is primitive recursive, by showing that

$$f(x) = \begin{cases} 0 & \text{if } x \text{ is odd} \\ 1 & \text{if } x \text{ is even} \end{cases}$$

is primitive recursive.

* 17.   Suppose F is a primitive recursive function, $F: N \rightarrow N$. Let G be the left inverse of F; that is, $G(F(n)) = n$. Is G primitive recursive? Explain.

P 18.   Show a flowchart for computing G of Problem 17.

P 19.   Let L be a programming language. We define the set of well-formed programs over functions, F, and predicates, P, as follows: $f \in F \Rightarrow f \in L$ and $x, y \in L, p \in P \Rightarrow$ are in L. Let $F = \{f_1: x \rightarrow x + 1, f_2: x \rightarrow x \dot{-} 1\}$, $P = \{p(x) \Leftrightarrow x \text{ is even}\}$, and domain $(x) =$ the set of natural numbers $= \{0, 1, 2, \ldots\}$.

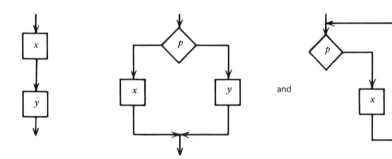

   (a) Is each function computable in L primitive recursive?

   (b) Let $g(x) = \begin{cases} 5 & \text{if } x = 0 \\ 0 & \text{if } x \neq 0 \end{cases}$. Is $x \leftarrow g(x)$ programmable in L?

P 20.   Let $(A, *)$ be a finite groupoid, where $A = \{0, 1, \ldots, k\}$. Show that the groupoid operation is primitive recursive. (Actually, your definition will then extend the groupoid operation, but you may ignore the extension.)

P 21.   Using the primitive recursive definition of addition, show that addition is associative.

P 22.   Show that if $f_1, f_2, \ldots, f_k$ are primitive recursive functions of n variables, then

$$f(x_1, \ldots, x_n) = f_1 \quad 0 \leq x_1 < n_1$$
$$f_2 \quad n_1 \leq x_1 < n_2$$
$$f_k \quad n_{k-1} \leq x_1$$

is primitive recursive.

P 23.   All recursive functions can be done in terms of alphabetical symbols. Specifically, let A be a finite alphabet. If x is a finite string over A, then $\sigma_a(x) = x \cdot a$. The base functions of this definition are:

Zero function: $n(x) = \lambda$ where λ is the null string.

Identity functions: U_i^n as usual.

Successor functions: $\sigma_a(x) = x \cdot a$ for each a in A.

Compositicn: as usual.

Primitive recursion: $h(x_1, \ldots, x_n, y) = f(x_1, \ldots, x_n)$

$g(x_1, \ldots, x_n, y \cdot a) = h_a(y, g(x_1, \ldots, x_n, y), x_1, \ldots, x_n)$

for each a in A.

Minimization: $\min_y \{g(x_1, \ldots, x_n, y) = \lambda\}$ where the set of strings over A is linearly ordered as follows: A is linearly ordered, $y < y \cdot a$ for every a in A, and strings of equal length are lexicographically ordered.

(a) Give a recursive definition, in terms of the preceding information, of predecessor.

predecessor $(y) = x \quad$ if $y = x \cdot a$
$\qquad\qquad\quad = \lambda \quad$ if $y = \lambda$

(b) Give a recursive definition, in terms of the above, of next.

next $(y) = y' \quad$ where $y' > y$
$\qquad\qquad\qquad$ and if $y'' > y \quad$ then $y'' \geq y'$

P 24. Let S be a sequence of n integers, $(S_1, S_2, \ldots, S_n)$ and θ a set of operations. For a sequence, T, of $n - 1$ choices from θ, $(\theta_1, \theta_2, \ldots, \theta_{n-1})$, we can write an expression:

$$E_T(S) = S_1\theta_1 S_2\theta_2 S_3 \ldots S_{n-1}\theta_{n-1}S_n$$

(a) Suppose that $\theta = \{+, -, \times\}$ with their usual precedences. Describe an algorithm to select the sequence T that will maximize $E_T(S)$.

(b) Apply your algorithm to the sequence

$$S = (3, 0, 4, -1, 5)$$

E 25. Give algorithms for the following.

(a) lcm (x, y) where lcm is the least common multiple.

(b) $\lfloor \sqrt{X \cdot Y} \rfloor$

E 26. Describe in detail (by giving the set of quintuples) a Turing machine that, starting on a tape with a finite string of 1's and 0's positioned at the leftmost nonblank, will produce a string of 1's and 0's that is the reverse of the original string and will stop with the head on the leftmost nonblank. [Rev $(0) = 0$, Rev $(1) = 1$, Rev $(x \cdot y) =$ Rev (y) Rev (x).]

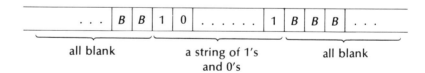

all blank   a string of 1's   all blank
and 0's

P 27. Suppose that in Problem 26 we restrict the input string so that it starts and ends with a 1. Interpreted as a binary number, this has some value: X. The result of reversing the binary representation of X is some number $f(X)$. f is a partial function because we have defined it for odd numbers only. Let g be any function that will act like f on odd numbers. Give an algorithm for a g of your choice.

E 28. (a) Can you simulate a Turing machine on a general purpose computer? Explain.

(b) Can you simulate a general purpose computer on a Turing machine? Explain.

E 29.   (a) Using truth tables, compute the function described by the propositional expression

$$(((p \vee q) \wedge \neg p) \supset (q \wedge \neg r))$$

(b) Show the tree of the expression of part a.
(c) Show the prefix form of the expression of part a.

E 30.   Determine the truth functional values of the following logical expressions.
(a) $((p \vee q) \supset ((p \vee r) \supset (r \vee q)))$
(b) $((p \supset ((q \vee r) \wedge s) \supset p) \equiv (\neg r \vee (p \wedge q)))$

P 31.   Determine logical expressions having the given truth tables of f and g.

p	q	r	$f(p, q, r)$	$g(p, q, r)$
T	T	T	F	T
T	T	F	T	F
T	F	T	T	F
T	F	F	F	T
F	T	T	T	F
F	T	F	F	T
F	F	T	F	T
F	F	F	F	T

P 32.   Derive an expression using only the connective $\{\supset, \neg\}$, which is truth-functionally equivalent to:

$$((p \wedge q) \wedge \neg(r \wedge q))$$

E 33.   Convert the following expressions first to tree form and then to prefix notation.
(a) $((p \supset (q \wedge r)) \supset (\neg r \vee (p \supset q)))$
(b) $((p \vee \neg \neg q) \supset \neg(q \wedge \neg(p \vee r)))$

E 34.   Using the notation where C, A, K, E, N are the prefix representations of $\wedge, \vee, \supset, \equiv, \neg$, respectively. Convert the following expressions first to tree form and then to infix notation.
(a) $CAKEpqNprs.$
(b) $CpAqKNpErs.$

P 35. Give a BNF description of the set of formulas in the prefix representation of the propositional calculus.

E 36. Define a new operation on logical variable by:

A	B	ψAB
T	T	F
T	F	T
F	T	T
F	F	T

Give formulas for $\vee AB$ and $\wedge AB$ using only ψ. (As an example, $\neg A$ is ψAA; you may use A or B more than once.)

P 37. Write a recursive program to go from standard form using the connectives $\neg$, $\vee$, $\wedge$, to a prefix form using only the ψ connective defined in Problem 36.

E 38. Develop a context-free grammar for the set of tautologies using only the connectives $(\neg, \supset)$ and the propositional symbols $\{a, b, c\}$.

P 39. Justify the principle of complementarity (Chapter 1, Section K) using structural induction. Do the same for the principle of duality.

SUGGESTED PROGRAMMING EXERCISES

1. Program a Turing machine interpreter and apply it to the program in Section D.

2. Program and run the translation problems in Section E to transform and evaluate logical expressions.

5

trees

SUMMARY

In this chapter we show how some of the ideas developed in earlier chapters can be applied to several of the building blocks used in the synthesis of programs. The tree is the first structure considered; directed, rooted, ordered, and finite trees are specialized, labeled digraphs and were introduced in Chapter 2.

Specializing from the class of all digraphs to trees allows us to develop tree representation and tree processing methods that generalize some better understood string processing algorithms. Thus we extend congruences and homomorphisms (Chapter 3) and finite automata (Chapter 2) to trees. Such generalizations are, in fact, one of the primary motivations for any formal study. In most cases, generalizations of results from strings to digraphs are not known and, in some cases, even the extension to trees has not been achieved.

Data structures that are more general than trees are briefly introduced. In addition to lists, which are discussed because their particular uniformity of structure makes implementation on conventional computers convenient, the more general concept of data structure is introduced with an indication of how some of the theory in Chapters 2 and 3 can be applied to these data structures.

A. TREES

A tree is a connected, acyclic, directed graph that has a distinguished vertex called the root; each vertex other than the root has in-degree 1, the root has in-degree 0, and the direct successor nodes of any node are linearly ordered. Nodes of the tree, which have no successors, are called *leaves*. The sequence of labels of the leaves from left to right is called the *yield* of the tree. Furthermore, all of the trees with which we are concerned are finite. In most cases, the nodes or vertices of the trees are labeled.

Example

The arithmetic expression $2 + 4 \times x - 5 \times y^2$ can be represented as the tree

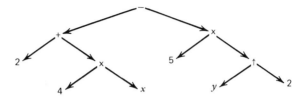

by fully parenthesizing the expression to the form

$$((2 + (4 \times x)) - (5 \times (y \uparrow 2)))$$

and by then representing the tree of each composite expression $(e_1 \theta e_2)$ as

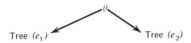

and each noncomposite expression as itself. Note that the root is shown at the top of the tree and all edges are directed downward. The yield of the tree is "24x5y2."   #

Since trees are acyclic, we adopt a convention similar to that used for Hasse diagrams, and we omit arrows from the edges of the graph. The root will always be drawn at the top of the diagram, and the directions of edges, if needed, can be inferred from the fact that for each vertex, v, in the tree, there is a unique path from the root to v.

The general usage of trees is to exhibit hierarchical structure. In this way, the tree of an arithmetic expression shows the structure of the arithmetic expression. Likewise, in a language specified by a context-free grammar, the structure of a sentence can be exhibited by its derivation tree.

Let $G = (V_N, V_T, S, P)$ be a context-free grammar. Then a single node labeled S is a derivation tree. Let T be a derivation tree, let x be a leaf of T labeled V, and let $V \rightarrow \alpha_1 \alpha_2, \ldots, \alpha_n$ be a rule in P. Then T' is a derivation tree, where T' is obtained from T by adding edges from x to $x_1, x_2, \ldots, x_n$ ordered in the natural number sequence from left to right and labeling the nodes $\alpha_1, \alpha_2, \ldots, \alpha_n$, respectively. A tree is a derivation tree *in the language*, $L(G)$, if the yield of the tree is a word in the language.

Example

Let $G = (V_N, V_T, S, P)$ with $V_N = \{S, T\}$, $V_T = \{a, b, c\}$, and $P = \{S \to aTb,\ T \to aSTb,\ T \to c\}$. Then the following is a derivation tree in $L(G)$

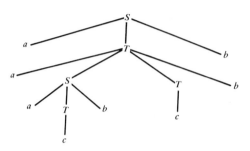

and the yield of the tree is *"aaacbcbb."*   #

If the variables of the context-free grammar have names whose semantics are apparent, this will be exhibited in the tree.

Example

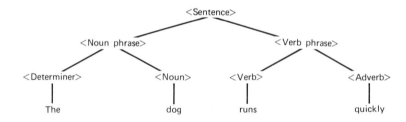

Example
    (Algol 60.)

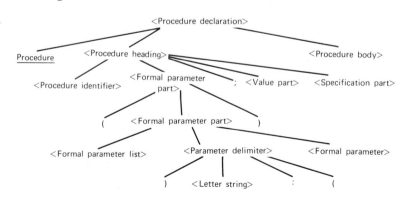

In order to discuss the representation of trees, the concept of a subtree must be defined. Let x be any node in a tree. Then the *subtree at* x is the set of all nodes reachable from x, $N(x)$, and all edges in $N(x) \times N(x)$. It is easy to see that the subtree at x is a tree justifying the name. If x is a leaf of the tree, then subtree at x is a trivial tree consisting of a single node. Note that a single node still satisfies the definition of a tree.

The direct successors of a given node x in a tree are ordered from left to right as $x_1, x_2, \ldots, x_k$. Each of these nodes, x_i, is the root of a tree, t_i. The trees $t_1, t_2, \ldots, t_k$ are called the *direct subtrees* of x.

The representation of a tree as a string of symbols uses the hierarchic structure of the tree. Thus the representations can be described as recursive processes. A typical example is the Polish prefix representation.

Definition

Given a tree, t, its *Polish prefix representation*, $P(t)$, is given as follows. (a) *If t consists of a single node, x, then $P(t)$ is the label of x; otherwise, (b) if $x = root$ (t) has successors $x_1, \ldots, x_k$ ordered from left to right, then $P(t) = label$ (x) $(P(t_1), P(t_2), \ldots, P(t_k))$.*

Example

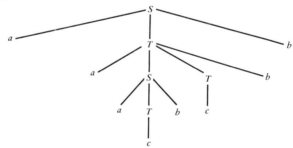

is represented as: $S(a(T(a, S(a, T(c), b), T(c), b), b)$.   #

If the alphabet used to label the tree is a ranked alphabet, so that the number of direct successor nodes of a node x can be determined uniquely from the label of x, the parentheses (and commas) are unnecessary and a parenthesis-free notation can be used.

Example

In the propositional calculus (Chapter 4) each propositional variable has rank 0, $\neg$ has rank 1, and each binary connective in $\{\supset, \&, \vee\}$ has rank 2. Since the rank of the symbols is uniquely determined, the parenthesis-free notation is, in the composite case, label (root) $P(t_1)P(t_2) \ldots P(t_k)$.

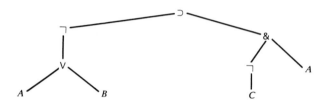

is represented as $\supset \neg \vee AB \& \neg CA.$ #

B. TREE DOMAINS

Gorn[1] described a tree as a Dewey decimal set of addresses, each address consisting of a sequence of 0 or more integers. Let λ denote the sequence of no integers. λ is taken to be the address of the root. Furthermore, if x is the address of any vertex, then $x \cdot 0, x \cdot 1, x \cdot 2, \ldots$ denote the addresses of the direct descendants of x from left to right.

Example

The following shows a tree whose vertices are labeled with their addresses:

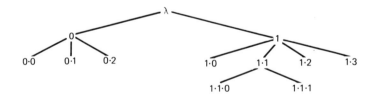

Definition

A tree domain, D, is a set of tree addresses that satisfies the following conditions.

[1] S. Gorn, "Explicit Definitions and Linguistic Dominoes," Proceedings of Conference on Systems and Computer Science, J. F. Hart and S. Takasu, eds., University of Toronto Press, 1965.

(a) *Each element of D is a sequence of integers.*

(b) *If x is in D and y is any prefix of x, then y is in D [y is said to be a prefix of x if x can be written as y · α, where α is a (possibly null) sequence of integers].*

(c) *If x is in D and x is of the form y · n, where n is a nonnegative integer, then each sequence of the form y · m, for 0 ≦ m ≦ n, is also in D.*

The tree domain definition provides a convenient abstract form of specifying a set of tree addresses. Given any finite set, we can easily check if it satisfies the conditions of the definition by determining whether the required elements are present for each member of the set.

Example

If $1 \cdot 2 \cdot 3$ is in D, then the following must also be in D: $\lambda, 0, 1,$ $1 \cdot 0, 1 \cdot 1, 1 \cdot 2, 1 \cdot 2 \cdot 0, 1 \cdot 2 \cdot 1, 1 \cdot 2 \cdot 2,$ and $1 \cdot 2 \cdot 3$. Any tree in which $1 \cdot 2 \cdot 3$ is an address must at least have the following.

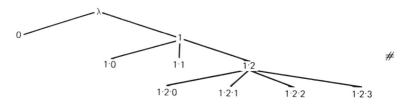

The *frontier* of a tree is the set of vertices having no direct descendants in the tree. If x is the address of a frontier vertex in the tree, t, with tree domain D, then x is not a proper prefix of any element of D.

Example

Let t be

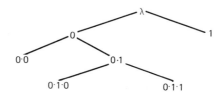

Then $D = \{\lambda, 0, 1, 0 \cdot 0, 0 \cdot 1, 0 \cdot 1 \cdot 0, 0 \cdot 1 \cdot 1\}$ and frontier $(D) = \{0 \cdot 0, 0 \cdot 1 \cdot 0, 0 \cdot 1 \cdot 1, 1\}$.   #

The prefix form of a tree is a listing with appropriate punctuation of the labels of the vertices of the tree. The order in which the vertex labels occur in the prefix form is the lexicographic order of the vertex addresses. Recall from Chapter 1 that the lexicographic order is defined as follows. Let α and β be sequences of integers, $\alpha < \beta$ if (a) α is a prefix of β; or (b) $\alpha = \alpha_1\alpha_2 \ldots \alpha_k$, $\beta = \beta_1\beta_2 \ldots \beta_l$, and $\alpha_i = \beta_i$, $1 \leq i < m$ and $\alpha_m < \beta_m$.

Example
    The lexicographic order of the vertex addresses of the tree of the previous example is:

$$\lambda, 0, 0 \cdot 0, 0 \cdot 1, 0 \cdot 1 \cdot 0, 0 \cdot 1 \cdot 1, 1 \quad \#$$

A *labeled tree* is a tree domain, D, and a mapping from D to a set of labels L. The *yield* of a labeled tree is the sequence of labels of the frontier vertices arranged in lexicographic order of their tree addresses.

Example

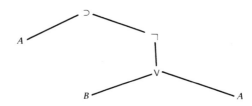

is a labeled tree. An unlabeled tree

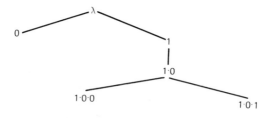

shows the arrangement of tree domain addresses. The tree domain, D, is given by

$$D = \{\lambda, 0, 1, 1 \cdot 0, 1 \cdot 0 \cdot 0, 1 \cdot 0 \cdot 1\}$$

and the labeling function is:

Address	Label
λ	$\supset$
0	A
1	$\neg$
$1 \cdot 0$	$\vee$
$1 \cdot 0 \cdot 0$	B
$1 \cdot 0 \cdot 1$	A

The frontier of this tree is, in lexicographic order $\{0, 1 \cdot 0 \cdot 0, 1 \cdot 0 \cdot 1\}$. The yield of the tree is ABA. The lexicographic order of D is $\lambda, 0, 1, 1 \cdot 0, 1 \cdot 0 \cdot 0, 1 \cdot 0 \cdot 1$, and the prefix form of the tree is $\supset A \neg \vee BA$.   #

The following is an alternative definition of finite trees. Let V be the set of vertices of the tree, and define two ordering relations on V.

1. *anc* is a partial order on V, where $x\,anc\,y$ means intuitively that x is an "ancestor" of y. If $x\,anc\,y$, the tree domain address for x is a prefix of the tree domain address of y.

2. *left* is transitive on V. If $x\,left\,y$ then, intuitively, if the tree were appropriately drawn, x would lie to the left of y.

A tree domain is defined axiomatically as a set of V of vertices with relations *anc* and *left*, such that:

1. For any m in V, the set $\{n \mid n\,anc\,m\}$ is well-ordered.

2. Any two vertices of V have a greatest lower bound under *anc*.

3. Any two distinct vertices are left-comparable if they are *anc*-incomparable.

4. If $m\,anc\,n$ and $m\,left\,p$, then $n\,left\,p$; if $m\,left\,n$ and $n\,anc\,p$, then $m\,left\,p$.

C. HOMOMORPHISMS AND CONGRUENCES ON TREES

We will assume in this section that the labeled trees are labeled from an alphabet, Σ, that is ranked. To each element $x \in \Sigma$ there is assigned a unique natural number through a ranking function, $r: \Sigma \to N$. If a node ν is labeled x, and if $r(x) = n$, then we require that ν have n direct successors.

Example
$$\Sigma = \{+, x, a\}$$

$$r: + \to 2$$
$$x \to 2$$
$$a \to 0$$

Then

is a tree labeled with elements of a ranked alphabet, Σ.   #

Each labeled tree represents, uniquely, a term from the *word algebra* over the same alphabet, where:

1.   Each element of rank 0 is a term.

2.   If θ is an element of rank k and $x_1, x_2, \ldots, x_k$ are terms, then $\theta(x_1, x_2, \ldots, x_k)$ is a term.

Since we are considering only ranked alphabets, the parentheses and commas may be omitted, and we can represent the term $\theta(x_1, x_2, \ldots, x_k)$ as $\theta x_1 x_2 \ldots x_k$.

A homomorphism, ϕ, of an algebra $A = (A, \Omega)$ to an algebra $B = (B, \Psi)$ maps $A \to B$ and n-ary operators in A to n-ary operators in B, where the condition on ϕ is that for any operators θ, η in $\Omega,$ Ψ respectively; $\phi(\theta(x_1, \ldots, x_n)) = \eta(\phi(x_1), \phi(x_2), \ldots, \phi(x_n)).$

Example
Let $A = (A, \Omega)$ be an arbitrary ranked alphabet and $B = (\{0\}, \Psi)$, where $\Psi = N - \{0\}$; and let $\phi: x \to \text{rank}(x)$ if x is an operator. Then each term is mapped to a sequence of integers.

Let $k = k_1k_2 \ldots k_n$ be the sequence of integers obtained from applying ϕ to a term in A. Then k is a term in B. A necessary condition on k is that, starting with the leftmost symbol, k_1, the sequence $k_2 \ldots k_n$ should be the concatenation of exactly k_1 terms. Then, if the sequence $k_ik_{i+1} \ldots k_n$ is the concatenation of r terms, the sequence $k_{i+1} \ldots k_n$ is the concatenation of $r + k_i - 1$ terms.

Let $G_1 = k_1$ and $G_{i+1} = G_i + k_{i+1} - 1$, $1 \leq i < n$, the necessary condition is $G_i > 0$, $i < n$, and $G_i = 0$ for $i = n$.   #

Example

The term $+x + yz$ is mapped to a sequence of ranks 20200, and $G = G_1G_2 \ldots G_5 = 21210$.   #

Recall the relationship between homomorphisms and congruences (Chapter 3) as contained in the following theorem.

THEOREM.   Let ϕ be a homomorphism of an algebra A onto an algebra A'. The equivalence relation, $E = \phi\phi^{-1}$, is a congruence on A, and there exists an isomorphism ρ of A/E onto A' such that $\phi = X \cdot \rho$, where X is the natural homomorphism of A onto A/E.

The diagram shown in Figure 5-1 is often helpful in visualizing the mappings involved. [Note that in Figure 5-1, ϕ^{-1} is usually not a function, but only a relation. Also, $E = \phi\phi^{-1}$ or, in other words, $aEb \Leftrightarrow a$ is in $\phi^{-1}(\phi(b))$.]

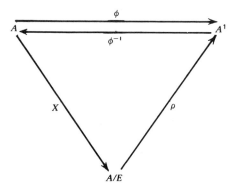

Figure 5-1.   Diagram relating the congruence, E, the homomorphism, ϕ, and the quotient algebra, A/E.

Example

Let p be the set of prefix form formulas (Chapter 4); then p is an algebraic system with the usual ranks of the operators; rank $(\supset)$ = rank $(\&)$ = rank $(\vee) = 2$, rank $(\neg) = 1$ and, if A is a propositional variable, rank $(A) = 0$.

Let ϕ be the homomorphism mapping symbol sequences in P to sequences of ranks. Then if x and y are two sequences such that $\phi(x) = \phi(y)$ then, for any u and v, uxv is a prefix form formula if and only if uyv is a prefix form formula.   #

Homomorphisms provide a convenient way of describing and programming the evaluation of formulas or expressions. The homomorphism in the evaluation of an expression is a mapping from the word algebra to a value algebra. Since the expressions in the word algebra are isomorphic to trees, the homomorphism is thus an evaluation of the tree.

Example

Let P be the set of prefix form formulas over the propositional variables A, B, and C. Let Q be the algebra of truth values, and let val be the homomorphism from P to Q that assigns truth values to A, B, and C. Then, if x is a formula, val (x) is the truth value of x.

Let $x = \supset\supset A \supset A \vee BC \neg A$ and val $(A) = T$, val $(B) =$ val $(C) = F$; then val $(x) = \supset\supset T \supset T \vee FF \neg T = T$.   #

Note that evaluation in the value algebra can be seen as a computation on a tree in the algebra of truth values. The values of the leaves of the trees are directly specified, and the value of any node in the tree is determined when the value of its direct successors is known. In this way, proceeding up the tree from the leaves to the root, the value of the whole expression can be determined.

Example

We continue our example of prefix form formulas with the tree representation of $\supset\supset A \supset A \vee BC \neg A$, which is:

The homomorphic image of this tree under truth value assignments is:

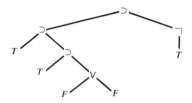

when val $(A) = T$, val $(B) = F$, and val $(C) = F$. The evaluation of this tree, using the recursive rule val $(\theta(x_1 \ldots x_n)) = f(\theta)$ (val $(x_1), \ldots$, val (x_n)), yields successively the following trees.

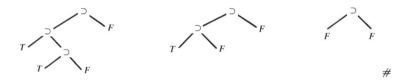

#

Example

As a last example of the evaluation of a tree using the homomorphic val, consider the arithmetic expression: $x + 5 \times (y - 3)$ whose tree is:

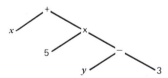

with val $(x) = 4$ and val $(y) = 7$. The homomorphic image of this tree is:

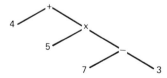

Evaluating this tree in bottom-to-top fashion yields the successive trees

#

D. AUTOMATA ON TREES

In this section, we discuss some of the results connected with tree automata. Essentially, a tree automaton is an automaton whose input is a tree from some set of trees. This generalizes conventional (finite) automata theory, where each input is a string (i.e., a very thin tree when written vertically) from some set of strings. We say that the domain of a tree automaton is a set of trees. The output of the tree automaton will either be (1) a "yes-no" answer, in which case we speak of a tree *acceptor* or *recognizer*; (2) a tree (or set of trees), in which case we call it a tree *transducer*; or (3) a string (or set of strings), in which case the automaton will be called a *translator*.

The primary reason for interest in tree automata is that trees play a central role in describing the phrase structure of sentences on the class of languages known as context-free languages. This interpretation has been carried over directly to programming languages in the use of the Backus-Naur form (BNF) as a generative grammar schema (i.e., a BNF grammar describes the syntax of a context-free language, and a generation of a word by a BNF grammar is representable as a derivation tree).

One of the objectives of tree automata is to extend in some "natural" way the body of results on finite automata (i.e., automata that operate on strings) to tree automata. Some of the results that one would like to carry over are a set of (tree) regular expressions, a minimization theory for tree automata, and a characterization of the formal properties of acceptors, transducers, and translators.

Several types of tree automata have been defined; some of them have direct analogs in string automata, and others seem to be new. We do not yet have a complete characterization of how these different automata are related, although some results are available and will be discussed.

Note first that in the case of finite automata, the automaton is described as a set of states, Q, a subset of which is final states, $Q_F \subseteq Q$, and a direct state transition function $M(\sigma, q) \to q'$, which describes the state change of the automaton in moving past the symbol σ, starting in state q. Because of the apparent symmetry of strings, we need not describe the direction of motion of the automaton, although, in fact, for any particular set of strings, the number of states does depend on the direction of motion.

In the case of trees (drawn with their root at the top), if the automaton is moving from the bottom up, we visualize a set of subautomata attached initially to the leaves of the tree. In the recognition process these subautomata move toward the top of the tree. At each node, the set of subautomata converging on the node is replaced by a single subautomaton that then moves upward.

When the automaton moves from the top down, we visualize an automaton splitting into autonomous subautomata at each node. These continue to move downward to the leaves.

In both cases, top-to-bottom and bottom-to-top recognition is definable in terms of the states of the automata when the process is completed. We will discuss only the bottom-to-top recognizers.

Definitions

Let Q be the state set, $Q = \{q_1, \ldots, q_n\}$, and let Σ be the alphabet that labels nodes of the trees.

The *transition function*, $M((q_1, \ldots, q_n), \sigma)$, describes the next state to appear at the node labeled σ when its direct descendants from left to right are in states $q_1, \ldots, q_n$.

An automaton is *deterministic* if the transition function $M(x)$ is a single state or the empty set of states (but not both) for each x in $Q^n X \Sigma$. Otherwise, the automaton is *nondeterministic*.

A set of trees R is *recognizable* if there is a tree automaton accepting exactly the trees in R.

A set of trees is *local* if it is the set of all derivation trees of a context-free grammar.

A set of trees T_1 is the *projection* of a set of trees T_2 if T_1 is obtained from T_2 by relabeling nodes through an onto mapping $\pi: \Sigma_2 \to \Sigma_1$, which changes alphabets. (T_2 is said to be the *inverse projection* of T_1 if it is the largest set of trees with labels in Σ_2 whose projection is T_1.)

The *yield* of a tree T is the concatenation of the labels of its leaves taken from left to right. The yield of a set of trees τ is the set of yields of the trees T in τ.

Example

We illustrate the recognition of trees corresponding to derivation in the grammar with productions.

$$S \rightarrow S_1 S_2$$
$$S_1 \rightarrow a S_1 b$$
$$S_1 \rightarrow ab$$
$$S_2 \rightarrow b S_2 a$$
$$S_2 \rightarrow ba$$

Clearly, $L(G) = \{a^n b^{n+m} a^m \mid n, m \geq 1\}$.

The transition function of the automaton is:

$$M(b) = q_1, M(a) = q_0$$
$$M((q_0, q_1), S_1) = q_2$$
$$M((q_1, q_0), S_2) = q_3$$
$$M((q_0, q_2, q_1), S_1) = q_2$$
$$M((q_1, q_3, q_0), S_2) = q_3$$
$$M((q_2, q_3), S) = q_4$$

The set of final states is $Q_F = \{q_4\}$.

The recognition of a tree by this tree automaton is illustrated in Figure 5-2.   #

Note that in each step of the recognition procedure a node x, all of whose descendants are labeled with state $\{q_i\}$, is replaced by a node labeled with a state.

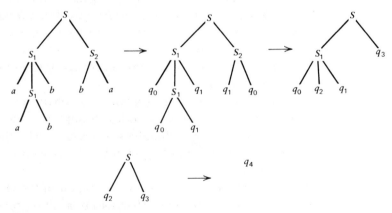

Figure 5-2.   Recognition by a bottom-up tree automaton.

It is clear that we need not restrict our tree automata to stratified alphabets, as seen in this example where both S_1 and S_2 are given two stratification numbers.

We summarize without proof some of the basic results of tree automata.

1. Whenever τ is the set of trees accepted by some tree automaton, the yield of τ is context-free. (Note that the yield of τ is a set of strings, and there is some context-free set of trees whose yield is the same as the yield of τ.)

2. Whenever τ is a local set, there is some tree automaton that recognizes τ. The converse is not true; there are sets of trees accepted by tree automata that are *not* local sets (i.e., generated by context-free grammars). (Further clarification of this is given by the following example and discussion.)

3. A set of trees is recognizable if and only if it is the projection of a local set.

4. A set of strings U is context-free (i.e., U is the yield of a local set) if and only if U is the yield of a recognizable set of trees.

5. Nondeterministic tree automata are no more powerful than deterministic tree automata.

6. The Boolean closure of recognizable sets of trees is recognizable (i.e., if τ_1 and τ_2 are recognizable sets, $\tau_1 \cap \tau_2$, $\tau_1 \cup \tau_2$, and $\neg \tau_1$ are recognizable). The complement may either be interpreted with respect to the set of all trees over the stratified alphabet if, indeed, the alphabet is stratified, or it may be interpreted with respect to some arbitrary, but finite, stratification relation. (Note that either of these interpretations is valid.)

7. Recognizable sets are closed under projection and inverse projection.

Example

To clarify some of the preceding summaries, consider the set of trees τ, corresponding to $S \to SS$; $S \to 01$; $S \to 10$; $S \to 00$; $S \to 11$ such that each tree in yield τ has an even number of 1's. τ is not a local set, but it is recognizable using the following automaton.

$$M(0) = q_0$$
$$M(1) = q_1$$
$$M((q_0, q_1), S) = M((q_1, q_0), S) = q_1$$
$$M((q_0, q_0), S) = M((q_1, q_1), S) = q_0$$
$$Q_F = \{q_0\}$$

τ is the projection of the local set of trees corresponding to:

$$S \to SS$$
$$S \to S_1 S_1$$
$$S_1 \to S_1 S$$
$$S_1 \to S S_1$$
$$S_1 \to 10$$
$$S_1 \to 01$$
$$S \to 00$$
$$S \to 11$$

under the projection $\pi: S \to S$; $S_1 \to S$; $1 \to 1$; $0 \to 0$.

The reason that τ is not a local set is that all of the rules of the grammar do occur in the derivation trees. Thus any tree construction from these rules must be in the local set determined by the grammar. The tree $\overset{S}{\underset{1 \diagup \diagdown 0}{}}$ is in the local set determined by this grammar, but the yield has an odd number of 1's. On the other hand, the rule $S \to 10$ is needed, since the

tree 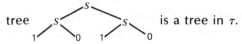 is a tree in τ.

The main point of this example is to clarify the distinction between a local and a recognizable set and the role played by tree automata. The tree automaton essentially is a syntax checker that checks the correctness of the rule applied at any given node. However, the tree automaton can also check a finite number of additional conditions, which allows it to accept trees that are context-free with an additional finite state condition. On the other hand, since the finite automata can only check finite *additional conditions*, this state information can be coded into the alphabet. Thus it is that the recognizable sets are projections of local sets.   #

A program to evaluate trees, from bottom to top, can be written directly as a recursive procedure from the tree automaton. A rule $M((q_i, q_j), S_k) = q_e$ is translated to:

> IF label (node) = S_k <u>and</u> value (left-descendant) = q_i
> > <u>and</u> value (right-descendant) = q_j
> > > THEN value (node) = q_e

while a rule $M(b) = q_i$ is translated to:

> IF label (node) = b <u>and</u> is-terminal (node)
> > THEN value (node) = q_i

The complete bottom-up tree automaton program is a sequence of conditional statements—one for each rule of the transition function.

E. TREE REWRITING SYSTEMS

In Chapter 4 phrase structure grammars were described as a method of presenting finite systems for generating sets (possibly infinite) of strings of symbols. In this section, we illustrate two different systems for generating sets of trees.

Definition

> *A regular tree grammar, G, is a rewriting system with a finite set of initial trees, Γ, over a stratified alphabet B of terminal and nonterminal symbols and a finite set of rewriting rules $\phi \to \psi$, where ϕ and ψ are trees over B. The production $\phi \to \psi$ is applied by rewriting ϕ as ψ where it occurs. The language generated by G is the (Polish prefix representation of the) set of trees over $A \subseteq B$, where A is a set of terminal symbols.*

Example

> Tree grammars have a superficial resemblance to grammars of the top-to-bottom variety, which rewrite trees. The produc-

tions in a tree grammar are of the form $T_1 \rightarrow T_2$ (i.e., a tree is replaced by another tree). Starting with the tree X and the tree rewriting rules $X \rightarrow a$ and $X \rightarrow$, we have the following derivation of

Note that the sequence of tree forms in the derivation is not unique, but corresponds to the leftmost derivation of a context-free grammar. At each step of the derivation, it is the leftmost nonterminal that is expanded.   #

The main result on regular tree grammars is that the language generated by any regular tree grammar is context-free. This is a special case of a semi-Thue system (i.e., type 0 grammar), which does not give rise to all recursively enumerable sets.

A set whose elements can be generated by an algorithm is called a *recursively enumerable* set. Consider the set of all Algol 60 programs. We can surely write a program that will generate all strings over a given alphabet. We do this by listing in lexicographic order all strings of length 1, then all strings of length 2, and so forth. In such a listing each Algol program will appear, but many non-Algol strings will also appear, including all of Shakespeare's plays, the Encyclopaedia Brittanica, and any other reasonable or nonsensical strings of symbols. As each output string is produced, we enter the string in a syntax checker that determines whether the string represents a valid Algol program.

If the string is a valid Algol program, the syntax checker adds it to the list of valid Algol programs. To this point, we have recursively enumerated the set of syntactically valid programs.

Next we ask for a list of all programs that will halt. (To simplify the discussion, assume that if a program ever requests input data, it then halts.) We know that Algol programs can simulate Turing machines, and so we cannot hope to determine whether a particular Algol program simulating a Turing machine will halt. However, we arrange for an execution machine capable of running Algol programs to take each program that is listed by the syntax checker and add it to an input list. The first execution machine will use the following routine. In its first cycle it will run the first program for 1 second. In its second cycle it will run its first two programs for 1 second each. In its kth cycle it will run the first k programs presented to it for 1 second each. Whenever a program that the execution machine is running halts, the execution machine will print it out.

In the course of events, any program that halts will ultimately be executed for enough seconds to complete and, therefore, will be listed. There will, of course, be some programs that will continue to be serviced by the execution program and will never halt. And, given any program, there is no algorithm to determine in which category it belongs—the set of programs that ultimately halts or the set of programs that goes on forever.

The set of programs that ultimately halts is, therefore, recursively enumerable. In fact, we have shown how, in principle, to enumerate them. The set of programs that never halts is not even recursively enumerable, but we will not show this.

If T is the set of trees generated by a regular tree grammar and ϕ is a set of tree rewriting rules, the set of trees obtained by applying ϕ to T is a regular tree language. In other words, the regular tree transformations are closed under composition. (A *regular tree transformation* is a set of trees R and a set of rewriting rules ϕ.)

Definition

A tree over a ranked alphabet *is a labeled tree in which each node labeled α has $r(\alpha)$ direct descendants, where r is the ranking function.*

The regular tree languages are related to the recognizable sets in the following ways.

1.   For every recognizable set R, there is a regular tree language T and a homomorphism (deleting some terminal nodes and the branches leading to them) such that $h(T) = R$.

2.   Every recognizable set over a ranked alphabet is a regular tree language.

An alternative style of description of tree sets is contained in the *tree adjunct grammars* or TAGs. We start with a description of tree adjunct grammars derived from context-free grammars.

A *minimal derivation tree* is a tree of depth 1, whose root is labeled with a nonterminal V and whose yield is the right side of a V-production. A *derivation tree* is defined recursively as either a minimal derivation tree or a tree obtained from two derivation trees T_1 and T_2 by replacing an occurrence of root (T_2) in the frontier of T_1 by T_2 itself.

Starting with the set of minimal derivation trees that corresponds to the rules of the grammar, we can recursively enumerate the set of derivation trees. The set of *basic derivation trees* is the finite subset of derivation trees such that on no path from the root to the frontier is any nonterminal duplicated (other than the root label, which may occur at most twice along a path); it includes precisely:

1.   The set of V-trees each of whose yields is $\Sigma^* V \Sigma^*$, called the set of *adjunct trees*, τ_a.

2.   The set of S-trees each of whose yields is in Σ^* and that has no interval nodes labeled S, called the set of *center trees*, τ_c.

Example
   We illustrate the set of basic derivation trees for the grammar, $G: \{S \to aTb; \ T \to TT; \ T \to ab\}$.

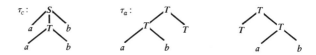

With the set of basic trees, we can describe the set of context-free derivations in G as follows.

(a) Every center tree is a derivation tree.

(b) If t is a derivation tree and t' is obtained from t by detaching the subtree t'' at some interval node ν, labeled V, attaching a V-tree in τ_a to ν, and attaching t'' to the V-node on the frontier of the adjunct tree, then t' is a derivation tree.

(c) Nothing else.

Example

See Figure 5-3.   #

It can be seen from the preceding development that the set of derivation trees in the TAG is exactly the set of derivation trees in the context-free grammar, and that for every context-free grammar, a TAG can be effectively derived.

By directly specifying a finite set of basic trees instead of deriving it from an underlying grammar, one can generate more than the context-free derivation tree sets. The resulting tree sets are incomparable with the recognizable sets.

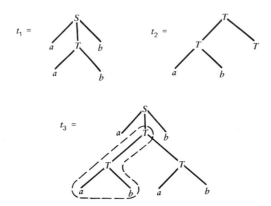

Figure 5-3.   A derivation step: t_3 obtained by adjoining t_2 to t_1.

F. SOME ADDITIONAL APPLICATIONS OF TREES

The Vienna Definition Language[2]

The Vienna Definition Language is an interpretive specification of the semantics of a programming language. It has been applied to give a formal definition of the semantics of PL/1.

Items in Vienna Definition Language are trees whose branches are labeled with *selectors* and whose terminal nodes are labeled with *objects* (that satisfy elementary predicates). The sequence of selectors tracing a path from the root to a node in the tree is called a *composite-selector*. The basic operation on trees is replacing a subtree at a selected node by another subtree. These tree operations are specified using the constructor, $\mu(x, y)$, where x is a tree and y is a set of (selector: replacement) pairs.

Example

If t is

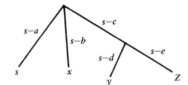

then $t^1 = \mu(t; \langle s \cdot b : s \cdot c(t) \rangle)$ is

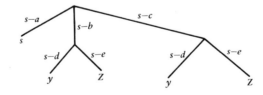

obtained from t by replacing the subtree at $s\text{-}b$ by the subtree at $s\text{-}c$.   #

---

[2]A tutorial exposition of the Vienna Definition Language is given in P. Wegner, "The Vienna Definition Language," *Computing Surveys*, 4 (1), 1972, pp. 5–63. For a more complete exposition, see P. Lucas and K. Walk, "On the Formal Description of PL/1," *Annual Review in Automatic Programming*, 6, Part 3, 1969, pp. 105–182.

The sequence of operations to be performed on the trees is specified by a recursive program. It is easy to show that the Vienna Definition Language has adequate computing power by showing that it can simulate a Turing machine.

Example

The structure of a Turing machine is given by the tree in Figure 5-4.   #

In Figure 5-4, s-state is a selector, and the object at the node reached from the root by following the state selector is specified to be a state by the elementary predicate is-state. The selector s-q provides us with a path to the listing of the "quintuples" that describe the actions of the Turing machine. If the Turing machine is in state σ_1, scanning symbol τ_1, on its tape, the selectors s-σ_1 and s-τ_1 will lead us to the node where the new state, new symbol, and head motions are specified. The s-tape selector leads to the description of the Turing machine tape, which is described as a tape to the left of the head, the symbol under the head, and the tape to the right of the head. Each tape half is described by the composite structure predicate:

is-Ω or

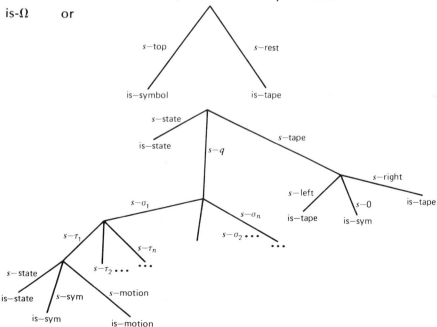

Figure 5-4.   VDL definition of Turing machine.

where is-Ω specifies any empty tape.[3] Note that the definition of is-tape is recursive; removing the top symbol from a nonempty tape still leaves an object satisfying the predicate is-tape.

First, consider the tape operations on the half-tape push-downs. To add a symbol to t (to push onto the pushdown), the operation is $\mu(t; \langle s\text{-top: is-symbol}\rangle\langle s\text{-rest: } t\rangle)$. To remove a symbol (i.e., to pop the pushdown), $\mu(t; \langle s\text{-top: } s\text{-rest} \cdot s\text{-top} (t)\rangle\langle s\text{-rest: } s\text{-rest} \cdot s\text{-rest } (t)\rangle)$, where the composite selector $s\text{-rest} \cdot s\text{-top } (t)$ is interpreted as $s\text{-top}(s\text{-rest } (t))$.

In the VDL program for the Turing machine, the usual VDL conventions are used: in the program, $A = B \rightarrow C$ is a conditional expression: if $A = B$ is true, then do C. In the sequence of program steps shown in indented form as

$$A$$
$$B$$
$$C$$
$$D$$

D is to be done first, followed by C, then B and, finally, A.

The program for running the VDL Turing machine is now given by:

exec $(t) = s$-state $(t) =$ halt $\rightarrow$ quit (t)
$\qquad T \rightarrow \underline{\text{exec}}\ (t)$
$\qquad\qquad \underline{\text{new-state}}\ (t)$
$\qquad\qquad\qquad \underline{\text{new-tapes}}\ (t)$
$\qquad\qquad\qquad\qquad \underline{\text{get-move}}\ (t)$

quit $(t) = \mu(t; \langle s\text{-state: } \Omega\rangle, \langle s\text{-}q: \Omega\rangle)$
get-move $(t) = \mu(t; \langle s\text{-move: } (s\text{-}q)(s\text{-}(s\text{-state}(t))) \cdot (s\text{-}((s\text{-tapes}) \cdot (s\text{-}0)(t)))(t)\rangle)$
new-state $(t) = \mu(t; \langle s\text{-state: } (s\text{-move}) \cdot (s\text{-state})(t)\rangle)$
new-tapes $(t) = (s\text{-move})(s\text{-action}(t) = \text{right} \rightarrow$
$\qquad\qquad \mu(t; \langle(s\text{-tapes}) \cdot (s\text{-left}) \cdot (s\text{-rest}): (s\text{-tapes})$
$\qquad\qquad\quad \cdot (s\text{-left})t)\rangle$
$\qquad\qquad \langle(s\text{-tapes}) \cdot (s\text{-left}) \cdot (s\text{-top}): (s\text{-move})$
$\qquad\qquad\quad \cdot (s\text{-sym})(t)\rangle$
$\qquad\qquad \langle(s\text{-tapes})(s\text{-}0): (s\text{-tapes}) \cdot (s\text{-right})(s\text{-}0)(t)\rangle$
$\qquad\qquad \langle(s\text{-tapes})(s\text{-right}): (s\text{-tapes})(s\text{-right})(s\text{-rest})(t)\rangle$
$\qquad\qquad$ —similar portions for left move or no move—

[3]Here we are, in effect, using a two pushdown store simulation of a Turing machine. Each half-tape is a pushdown.

The program *exec* is run. If the state of the Turing machine is a halt state, then the *quit* program is invoked, which erases the state and quintuple information and leaves only the tape as the result. Otherwise, the next move of the Turing machine is begun by the subprogram *get-move*, which creates a new selector branch, *s-move*, and attaches to it the information of the new-state, new-symbol, and head motion. After this is done, new-tapes "repositions the head on the tape." In the case shown—a move to the right—the left portion of the tape has the symbol just printed by the Turing machine "pushed" onto it, the top symbol of the right tape moves under the head, and the top symbol is removed from the right side of the tape. After the tape has been updated, the state information of the move replaces the old state information. Finally, *exec* is called again to repeat the cycle.

Transformational Grammar

Another application of trees is in the specification of the grammar of a natural language, such as English. A brief overview of some of the main components of such a transformational grammar follows.

1. The grammar consists of a context-free base of trees. The trees are the set of derivation trees of a context-free grammar. The terminal strings derived by the base grammar are not necessarily sentences in the language being described, but the context-free derivation tree is a representation of the deep-structure of the sentence to be derived.

2. There is a set of transformation rules, mapping trees to trees, that changes the structure of trees so that the yield of the tree, which results at the end of a sequence of transformations, is a sentence in the language. Transformations can rearrange the sequence of subtrees, adding or deleting components. A typical transformation is the transformation that converts a sentence from declarative form to passive form, interchanging the positions of subject and object and adding a form of the verb "to be" which changes the verb to a participle, and adding the preposition "by." Thus "Sharon ate dinner" becomes "Dinner was eaten by Sharon." Even in this simple case, a description of the transformation must refer to the grammatical categories of

words in the sentence and, hence, implicitly to the tree structure.

3. The transformational rules are stated as an *applicability condition* and a *structural change statement*. The applicability condition is a predicate that is evaluated over the tree to determine if the structural change should be applied.

In addition, the formulation of a transformational grammar generally imposes an ordering on the rules so that certain rules must be applied before others. Additional constraints can be added to give control over the sequence of applied rules to an arbitrary program.

Example

As an example of transformations in a transformation grammar, consider the context-free rules of the base:

$$S \rightarrow NP\text{-}VP \qquad\qquad det \rightarrow the$$
$$NP \rightarrow det\text{-}N \qquad\qquad det \rightarrow a$$
$$N \rightarrow Adj\text{-}N \qquad\qquad N \rightarrow dog$$
$$\qquad\qquad\qquad\qquad N \rightarrow cat$$
$$VP \rightarrow (Adv)\text{-}V\text{-}NP \qquad N \rightarrow mouse$$
$$Adj \rightarrow huge \qquad\qquad Adv \rightarrow quickly$$
$$Adj \rightarrow small \qquad\qquad Adv \rightarrow eagerly$$
$$V \rightarrow chased \qquad\qquad V \rightarrow ate$$

In this grammar we can derive the sentence "The small cat chased the huge dog" as follows.

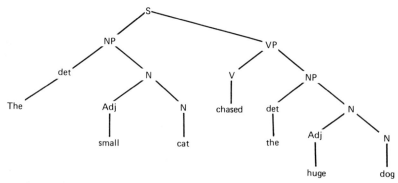

A transformational rule to change this sentence to passive will take the form:

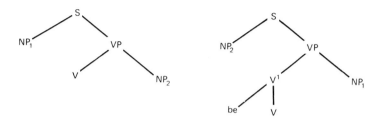

The resulting passive sentence tree structure will be:

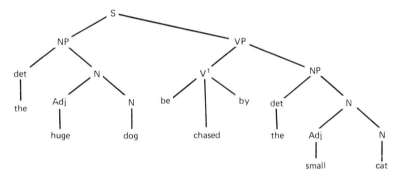

When rules are added to modify the tense of "be," the resulting sentence is "The huge dog was chased by the small cat." Notice that any modifying phrases attached to a noun phrase must move with the noun phrase in the transformation; these modifying phrases may have an arbitrarily deep tree structure and may displace the sentence components an arbitrary distance in the declarative form of the sentence, as in: "The small cat that was owned by Farmer Jones who had just moved to the suburbs after a productive career teaching in the agricultural school of the University of Delaware chased the huge dog." The corresponding passive form is: "The huge dog was chased by the small cat that . . . ."

G. LISTS

A list is a representation of data in memory where connections between items are shown explicitly by *pointers*. A pointer is a data type that can store a memory address.

The explicit representation of data connections is valuable because of the great flexibility it allows in possible data relationships. The price for this amount of flexibility is that in the most general usage of lists, the operations must be programmed explicitly by the user and are not provided by the system.

It is usual to represent list structures by means of diagrams in which a rectangle represents a unit cell of memory, an arrow is a pointer, and a cell with a diagonal line through it represents a null pointer.

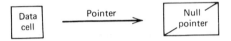

The representation of a sequence (a_0, a_1, a_2, a_3) can then be shown as:

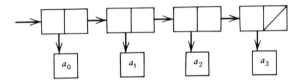

In this representation of the sequence, each position is associated with a pair of pointers: a pointer to the data item of the cell, and a pointer to its successor in the list.

Additional links can be added to this data representation. For example, each item can also have a pointer to its predecessor as well as to its successor. In this case, the list (a_0, a_1, a_2, a_3) could be represented as:

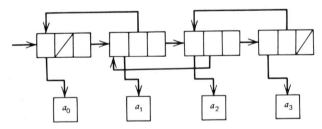

In the two-way list representation the typical cell of the list is:

Pointer to item	Back pointer	Forward pointer

The implementation of lists in a programming language

depends on the primitive operations that are available. If the programming language has facilities for pointers, such as PL/1, these can be used directly. If the programming language has indirect addressing, such as typical assembly languages, pointers are easy to implement. Even in a programming language such as Fortran or APL, list processing can be incorporated. In this case an array of two dimensions can be used with a row of an array comprising a composite data cell.

Example
   Representation of (a_0, a_1, a_2, a_3) in a two-dimensional array is as follows.

1	5	2
2	6	3
3	7	4
4	8	0
5	a_0	0
6	a_1	0
7	a_2	0
8	a_3	0

Elements $A(i, 1)$ $1 \leq i \leq 4$ are pointers to the data, while elements $A(i, 2)$ $1 \leq i \leq 4$ are pointers to successor cells in the list.   #
   List processing languages such as Lisp have implicit pointers and explicit operations for building lists. In this case, since the pointers are not explicit, the types of lists built may be more restricted. Let the typical list cell be a pair of atomic cells, Head (x), the first cell of the pair, and Tail (x), the second cell of the pair. The basic list building primitive operator is CONS. Assuming that NULL is a name for the null pointer, then

CONS $(A, NULL)$ will construct the cell ⟶ [A /]. Similarly,

CONS$(A,$ CONS $(B,$ NULL$))$ constructs the list.

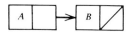

Example
   The representation of a function as a list of ordered pairs $(d_1, r_1), (d_2, r_2), \ldots, (d_n, r_n)$ could be arranged as:

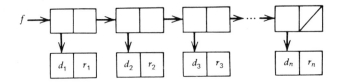

The search for r_k can then be programmed as:

Temp ← f;
WHILE HEAD (HEAD(Temp)) ≠ d_k DO
        Temp ← TAIL (Temp) DO;
Answer ← TAIL (HEAD(Temp))   #

In addition to the basic construction operators, such as CONS, and selection operators, such as HEAD and TAIL, list processing languages typically include predicates for identifying atomic data, the null pointer, and control structures such as sequential, conditional, recursive, and subroutine.

H. STRUCTURAL INDUCTION ON LISP-LIKE LISTS

The class of Lisp-like lists[4] can be defined by the context-free specification:

$$l_0 \to a_1|a_2| \ldots |a_n$$
$$l_1 \to \lambda|Cl_0l_1$$

where elements of the syntactic category l_0 are called atoms, and elements of the syntactic category l_1 are called linear lists. The symbol λ denotes the element called NIL.

An operator θ taking operands of type $\alpha_1, \ldots, \alpha_n$, respectively, and producing a result of type β is said to be of type $\beta/\alpha_1 \ldots \alpha_n$. (This is, of course, just the kind of heterogeneous algebra referred to at the beginning of Chapter 3, where the carrier of the heterogeneous algebra is partitioned into different classes, which are called types here.) Using this notation, C is seen to be of type l_1/l_0l_1, since the operands of C are of type l_0 and l_1, respectively, and the result is of type l_1.

---

[4]The name is derived from the programming language Lisp.

Next, we define two operators as follows. K is of type l_1/l_1l_1. K is a concatenation of lists. Since we are dealing with linear lists, each list can be represented as a finite sequence. If neither of the two operands of K is λ, the two operands can be represented as $L_1 = (a_1, \ldots, a_m)$ and $L_2 = (b_1, \ldots, b_n)$. KL_1L_2 is then $(a_1, \ldots, a_m, b_1, \ldots, b_n)$. λ is an identity element for concatenation. However, the formal definition for K is given by the following pair of tree transformation rules.

rule 1

$$
\begin{array}{c}
\quad\;\; K \\
\;\; C \quad\; Z \\
X \quad Y
\end{array}
\quad\longleftrightarrow\quad
\begin{array}{c}
\quad\;\; C \\
X \quad\; K \\
\quad Y \quad Z
\end{array}
$$

rule 2

$$
\begin{array}{c}
\;\; K \\
\lambda \quad Z
\end{array}
\quad\longleftrightarrow\quad Z
$$

where X is of type l_0, and Y and Z are of type l_1.
    M is of type

$$
\frac{l_0}{(l_0/l_0l_0)\,l_1l_0}
$$

M is known as MAP and applies a function to a linear list and an atom to obtain an atom. $M(+ (2, 3, 4)1) = 10$, while $M(\times (5, 4, 3, 2)1) = 5!$ Again, the formal definition of M is given by a pair of tree transformation rules.

rule 3

$$
\begin{array}{c}
\quad\quad M \\
f \quad C \quad z \\
\quad x \quad y
\end{array}
\quad\longleftrightarrow\quad
\begin{array}{c}
\quad\quad f \\
x \quad\; M \\
\quad f \quad y \quad z
\end{array}
$$

rule 4

$$
\begin{array}{c}
\quad M \\
f \;\; \lambda \;\; z
\end{array}
\quad\longleftrightarrow\quad z
$$

Both functions M and K are usually defined in linear notation, since their definitions are to be used as objects in a programming language. (Write linearized versions of these definitions in your favorate recursive programming language.) We can now prove the following theorem.

THEOREM

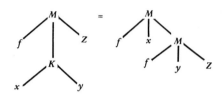

The proof proceeds by cases on x. Since x is of type l_1, it is either λ or is of the form cl_0l_1.

Case I. $x = \lambda$.

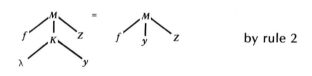
                                                                    by rule 2

while

                                                                    by rule 4.

Case II. $x = \overset{c}{\underset{v \quad u}{\diagup \diagdown}}$.   Here we need an induction principle for structures similar to the induction principle for trees, whose structural induction was introduced in Chapter 4. The induction principle (for structures) states that, for some set of structures, a structure has a certain property whenever all its proper constituents have that property, then all the structures in the set have that property. The induction principle on trees essentially states that if assuming a property of every tree of depth less than k implies that is true of depth k, then it is true of every tree. We assume the lemma is true for depth less than k and prove it for depth k.

Although many such proofs are "linearized" using linear list notation for trees, we feel that the two-dimensional tree notation is an aid in understanding all tree processing and transformations.

I. DATA STRUCTURE

A data structure is most generally a formal means for representing data relationships in the memory of the computer. A set of data objects represented in memory can be combined through a set of programmed operations. The programmed operations and the set of data objects are thus an algebra of data structures—*if* they satisfy the conditions described in Chapter 3. (Otherwise, their description may be provided by a partial, heterogeneous algebra.) Different representations of data objects constitute different algebras, which may be related through the fundamental theorems for mapping between algebras. The operators in such an algebra of data structures are often referred to as *constructors*.

The description of the data objects and the associated operations in a computational framework will utilize the mechanism of formal systems described in Chapter 4. Thus a computational algebra may be specified as a grammar or other suitable formal system. Of course, all such programming will be subject to the limits of the algorithmic method, which are implicit in the unsolvability of the halting problem. For example, the equivalence of independent descriptions of sets of objects may be undecidable or a computation on a data structure may be nonterminating.

REVIEW

The application of the algebraic and formal concepts to the building blocks of programs has been introduced with emphasis on trees, where the theory is perhaps better understood, instead of on arbitrary digraphs. Of course, an additional motivation for concentrating on trees is to simplify notations.

PROBLEMS

P   1.   A tree is a *binary tree* if each nonterminal vertex of the tree has exactly two successors. Derive a formula for the number of binary trees, $B(n)$, as a function of the number of nonterminal vertices, n.

E   2.   A *list-represented tree* is connected, directed, ordered graph having a distinguished root vertex; each vertex other than the root has indegree 1, the root has indegree 0. Each vertex has two potential direct successors—LEFT and RIGHT. The set of trees can be mapped 1-1 onto the set of list-represented trees as follows.

If $t = $ is a tree with root labels x and direct

subtrees $t_1, t_2, \ldots, t_n$, the list-represented tree, $L(t)$ is

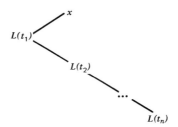

Example

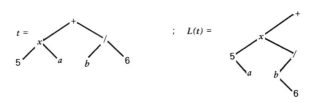

(a) Show that the mapping is 1-1.

(b) If the sequence of trees $(t_1, t_2, \ldots, t_n)$ is represented as the list-represented tree

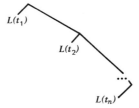

show that the mapping from sequences of trees to list-represented trees is 1-1, and onto.

P 3. Develop the definition of suffix representation of trees.

E 4. Describe algorithms for converting arithmetic expressions from infix to prefix form, and vice versa.

E 5. Modify the algorithms of Problem 4 to convert between infix and suffix form representations.

E 6. The suffix representation of an arithmetic expression is useful in computing the program of a stack machine. (A stack is a last-in, first-out arrangement of memory with data being added to and removed directly from the top of the stack.) In such a stack execution, the suffix representation is interpreted as a sequence of operations on the stack. The meaning of an operand x is "add x to the stack." The meaning of an n-ary operator, θ, is "removes the top most n values from the stack, $x_1, x_2, \ldots, x_n$ (where x_1 is the top of the stack) and add $\theta(x_1, x_2, \ldots, x_n)$ to the stack." Thus the sequence of stack configurations in evaluating $(5+4)x(3+6)$, whose suffix is $54+36+x$, is:

5		
5	4	
9		
9	3	
9	3	6
9	9	
81		

Using commutativity and associativity of operators, it is often possible to rearrange the tree of the arithmetic expression to require fewer stack positions in evaluating an expression.

(a) For the tree

$t =$

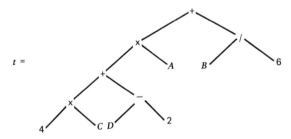

give the suffix form of the tree and compute the stack depth required to evaluate it.

(b) Rearrange the tree so that you can evaluate it with as shallow a stack as possible.

(c) What general principles of tree rearrangement can you deduce to minimize stack depth in evaluation?

E 7. Express the tree of Problem 6 as a mapping from a tree domain to a set of labels.

P 8. In the list-represented trees (Problem 2) define a tree-addressing notation as follows. The address of the root is λ; if x is the address of a node, then $x \cdot 0$ is the address of its LEFT successor, and $x \cdot 1$ is the address of its RIGHT successor.

Example

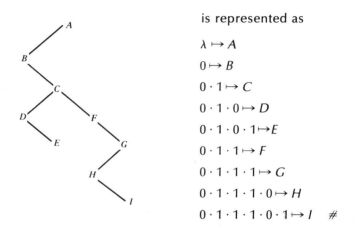

is represented as

$\lambda \mapsto A$

$0 \mapsto B$

$0 \cdot 1 \mapsto C$

$0 \cdot 1 \cdot 0 \mapsto D$

$0 \cdot 1 \cdot 0 \cdot 1 \mapsto E$

$0 \cdot 1 \cdot 1 \mapsto F$

$0 \cdot 1 \cdot 1 \cdot 1 \mapsto G$

$0 \cdot 1 \cdot 1 \cdot 1 \cdot 0 \mapsto H$

$0 \cdot 1 \cdot 1 \cdot 1 \cdot 0 \cdot 1 \mapsto I$ #

(a) Give a definition of tree domain for list-represented trees.

(b) Let $f: n \to 0 \cdot 1 \cdot 1 \ldots 1$ be extended to strings.
n1's
Show that in the list-represented tree corresponding to a tree, the node whose address is x, is mapped to the node whose address is $f(x)$.

P 9. Write a program to check if D is a tree domain.

P 10. Show that if t is a tree under the tree domain definition then, under the interpretation of *anc* and *left*, t satisfies the axioms given in Section B.

* 11. (a) Show that if t has the two ordering relations *anc* and *left*, which satisfy the axioms given in Section B, then the tree domain as a set of vertices can be defined.

(b) Using Problem 10 and part a, show that the definitions are equivalent.

E 12. In APL, arithmetic expressions are represented in infix form and are evaluated strictly from right to left. Describe algorithms to convert an arithmetic tree to an APL expression, and vice versa.

E 13. Draw the tree of

$$\supset \neg \supset A \vee BA \supset \& \supset BAAC$$

and evaluate it when A, B, and C are all *true*. Is this expression a tautology? Justify your answer.

* 14. Let T be the set of tautologies over the propositional formulas with propositional variables $\{A, B\}$. Give a type 2 grammar for T.

* 15. Consider the set of prefix form formulas in the propositional calculus. Is the set of tautologies a congruence class? Explain.

P 16. Let T be the set trees of formulas in the propositional logic. Describe a bottom-to-top tree automaton to evaluate these formulas.

E 17. (a) Let T be the set of trees of arithmetic expressions over the integers with operators $\{+, -, \times\}$. Describe a bottom-to-top tree automaton to compute the remainder of the value of the expression represented by the tree modulo 9.
(b) Explain why a finite tree automaton cannot evaluate arithmetic expressions.

E 18. Let T_1 be the following set of trees. Each nonterminal node is of rank 2 and is labeled with A, B, or C, except that the direct successor of any node, x, is not labeled the same as x. Each terminal node is labeled d or e. T_2 is the subset of T_1, defined by

$$T_2 = \{t \mid t \in T_1 \quad \text{and yield} \quad (t) \in d^*e^*\}$$

Is T_2 recognizable? Is T_2 local?

SUGGESTED PROGRAMMING EXERCISES

Problem 9 is a programming problem. So are the following exercises.

1. Write programs to convert a tree over a ranked alphabet from prefix form to a function on a tree domain, and vice versa.

2. Program a bottom-up tree automaton to recognize derivation trees of the following context-free grammar.

$$E \rightarrow E + T$$
$$E \rightarrow T$$
$$T \rightarrow T * F$$
$$T \rightarrow F$$
$$F \rightarrow (E)$$
$$F \rightarrow a$$

3. (a) Modify your program of Problem 2 to evaluate the trees when "a" is replaced by an integer, $0 \leq o < k$, and $+$ and $*$ are interpreted as modulo k operations.
    (b) Do the same as in part a, but $+$ and $*$ are the usual arithmetic operations. (Note that this program no longer corresponds to a finite state tree automaton.)

programming applications

SUMMARY

Computer science may be defined as the science of programming, where we understand the science to encompass the necessary theory and some of its applications. This definition, of course, parallels the usual definitions of the older sciences; we might explain chemistry as the science of the composition of substances and their transformations, and it is understood to include much of the theory of the electrical structure of atoms and molecules.

In this chapter, we give two examples of the application of the methods of discrete structures in the development of programs. Hopefully, this will illustrate further the utility of these methods.

Our first example describes, from the algebraic point of view, the incorporation of abstract structures in a concrete program that may not directly have the necessary primitive algebraic operations.

Our second example is a fragment of a compiler. Again, abstract structures must be somehow represented internally but, since this example is much more complicated, the process is decomposed into a sequence of transformations.

A. DATA STRUCTURES

The importance of abstraction in computer science was introduced in Chapter 0; in subsequent chapters different mathematical abstractions pertinent to computer science were introduced (Chapter 1, sets and functions, Chapter 2, graphs, and Chapter 3, algebraic systems). We know also from Chapter 4 that the computational model used to describe the abstraction must correspond in a computable way with the abstraction. Otherwise, the question of whether the abstraction and its supposed computational equivalent yield the same results may be algorithmically undecidable. For example, it is undecidable whether a given flowchart, F, and a given program, P, specify the same function. Only if P is derived algorithmically from F can we be assured that P is a correct implementation of F.

The development of an algorithm to solve a problem is done at a level where the necessary abstractions are available or may be introduced. At the first stage of algorithm development, one may talk of sets, graphs, trees, or algebraic structures. Later, in preparing the algorithm to be run in some computer language, one must translate the abstract structures and operations to the available primitives of the language.

Example

Let t be the class of binary trees, where trees are ordered and labeled as in Chapter 5. We can define the class of trees recursively as follows.

(a) A node labeled x is a binary tree, where $x \in A$.

(b) If t_1 and t_2 are binary trees, then 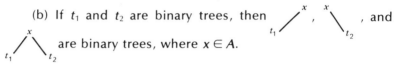, and

are binary trees, where $x \in A$.

(c) Nothing else is a binary tree.

If x is a finite alphabet, the set of prefix representations of binary trees is a context-free language. In a language such as Fortran, which does not include pointers, one might use an array structure to implement binary trees. Let A of dimension $n \times 3$ be used for a binary tree with n nodes with $A[i, 1]$ used to store the value of ith node, $A[i, 2]$ the index of the left successor of node i, and $A[i, 3]$ the index of the right successor of node i. The index of the root is 1 and, if node i has no left successor, $A[i, 2]$ is 0; if node i has no right successor, $A[i, 3]$ is 0.

The following tree, t, and an array form for t illustrate the implementation.

A	1	2	3
1	a	2	3
2	b	0	0
3	c	4	5
4	d	0	6
5	e	0	0
6	f	0	0

Note that the correspondence between trees and arrays has been defined in such a way that for a given tree, a unique array is not specified (although the tree corresponding to an array *is* unique). We, therefore, add the requirement that the nodes in the array appear in their prefix order (Chapter 4). The resulting array A' follows.

A'	1	2	3
1	a	2	3
2	b	0	0
3	c	4	6
4	d	0	5
5	f	0	0
6	e	0	0

Denoting by C the *coding mapping* from trees to arrays, we see that C is an injective function. Denoting by R the *representation (partial) function* from arrays to trees, R is defined only for arrays that are the representations of trees. Furthermore, R and C are inverses, in that $C \circ R$ is the identity on trees, and $R \circ C$, where defined, is an identity on arrays.   #

Any structure that is to be used as an abstraction in computation must be embodied in a computational equivalent. Calling these structures the *abstract* and *concrete* structures, respectively, the (partial) functions relating them are as shown.

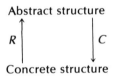

In terms of this correspondence between the abstract and concrete structures, one can specify the computation on the abstract structure, as shown in the following commutative diagram.

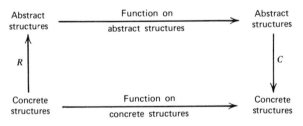

Example

We can represent (labeled, directed, ordered) trees as binary trees, using the rules that if t is a tree and $b(t)$ is the corresponding binary tree:

(a) The leftmost direct successor of x in t is the left successor of x in $b(t)$.

(b) The right sibling of x in t is the right successor of x in $b(t)$.

Thus $t =$ corresponds to $b(t) =$

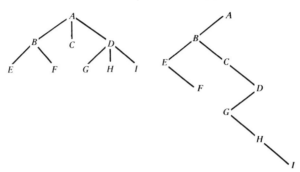

Now the coding of t to $b(t)$ may be composed with the coding of $b(t)$ in an array, which yields the following representation.

ARRAY REPRESENTATION OF t.

	Label	Left Successor	Right Successor
1	A	2	0
2	B	3	5
3	E	0	4
4	F	0	0
5	C	0	6
6	D	7	0
7	G	0	8
8	H	0	9
9	I	0	0

Note that since prefix order is preserved by each of the coding functions, $t \to b(t)$ and $b(t) \to$ array (t), it is preserved in the composite coding function $t \to$ array (t). Thus the successive node labels of the tree in the array yield the prefix listing of the tree.

Evaluation of the tree can be done entirely in terms of the array. If x is a node with left successor y and right successor z, the value of x is the list $x(y), z$. Starting at the highest index of the array, we may evaluate the array.

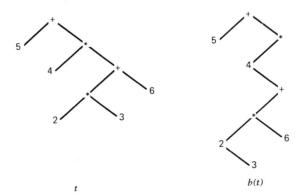

t

$b(t)$

ARRAY OF t WITH EVALUATION FROM HIGHEST INDEX TO LOWEST.

	Label	Left Successor	Right Successor	Value
1	+	2	0	$+ (5, 48) = 53$
2	5	0	3	5, 48
3	*	4	0	$*(4, 12) = 48$
4	4	0	5	4, 12
5	+	6	0	$+ (6, 6) = 12$
6	*	7	9	$*(2, 3), 6 = 6, 6$
7	2	0	8	2, 3
8	3	0	0	3
9	6	0	0	6

#

B. AN ALGEBRAIC STYLE CORRECTNESS PROOF FOR A SIMPLE COMPILER[1]

The algebraic point of view in relating different data structures is also applicable to compilation; the source language and its associated semantics can be understood as one algebraic system, and the object language and its associated semantics is another algebraic system. Compilation can then be seen as a means of relating algebraic systems.

[1]The material in the rest of the chapter is based on R. M. Burstall and P. J. Landin, "Programs and their Proofs: An Algebraic Approach," *Machine Intelligence*, 4, 1970, pp. 17–43.

The source language consists of the natural numbers $0, 1, 2, \ldots$ with the operations of addition and multiplication, and a set of variables $u, v, w, \ldots$. The syntax of expressions is:

1. $\langle \text{expression} \rangle \rightarrow \langle \text{expression} \rangle + \langle \text{term} \rangle$

2. $\langle \text{expression} \rangle \rightarrow \langle \text{term} \rangle$

3. $\langle \text{term} \rangle \rightarrow \langle \text{term} \rangle \times \langle \text{factor} \rangle$

4. $\langle \text{term} \rangle \rightarrow \langle \text{factor} \rangle$

5. $\langle \text{factor} \rangle \rightarrow (\langle \text{expression} \rangle)$

6. $\langle \text{factor} \rangle \rightarrow \langle \text{variable} \rangle$

7. $\langle \text{factor} \rangle \rightarrow \langle \text{constant} \rangle$

The semantics of expressions is:

1. $v(\langle \text{expression} \rangle) = v(\langle \text{expression} \rangle) + v(\langle \text{term} \rangle)$

2. $v(\langle \text{expression} \rangle) = v(\langle \text{term} \rangle)$

3. $v(\langle \text{term} \rangle) = v(\langle \text{term} \rangle) \times v(\langle \text{factor} \rangle)$

4. $v(\langle \text{term} \rangle) = v(\langle \text{factor} \rangle)$

5. $v(\langle \text{factor} \rangle) = v(\langle \text{expression} \rangle)$

6. $v(\langle \text{factor} \rangle) = v(\langle \text{variable} \rangle)$

7. $v(\langle \text{factor} \rangle) = v(\langle \text{constant} \rangle)$

For each rule in the syntax there is a corresponding semantic rule for evaluating the left side in terms of the right side components. Several points to notice are:

1. The operator symbols in the syntax are just symbols. In the semantic part the operator symbols indicate that the operation is to be performed.

2. When the derivation tree is exhibited, the evaluation proceeds bottom-up. The values of the constants are the numbers that correspond to the numerals in the expressions, and the values of the variables are the values of numerals associated with those values.

Example

$3 \times (u + 5) + v$, where u is 2 and v is 6. The derivation tree, using E, T, and F for expression, term, and factor, respectively, is shown in Figure 6-1. The values associated with the nodes in Figure 6-1 are shown in square brackets.   #

The compilation of expressions will proceed through a sequence of intermediate compilations to machines, so that the composition of all the intermediate compilations is the desired result. The sequence of intermediate machines is: (1) a stack machine; (2) a store-pointer machine on which the stack is

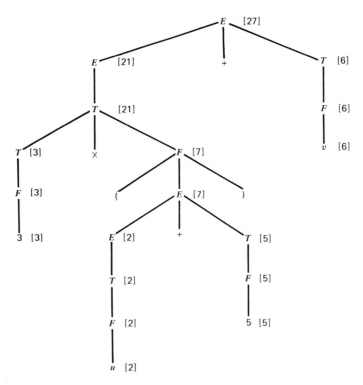

Figure 6-1.   Derivation tree for $3 \times (u + 5) + v$.

represented by a sequence of contiguous memory locations and a pointer is retained to address the top of the stack; (3) an address-program machine in which the stack data are retained in memory in the same way as the store-pointer machine, but the pointer information is incorporated into the instruction stream; and (4) a conventional store-accumulator machine. Table 6-1 shows the program for $3 \times (u + 5) + v$ for each of these machines and the corresponding memory configuration after each instruction.

TABLE 6-1*a* THE STACK MACHINE.

Program	Stack Sequence Empty
3	3
u	2, 3
5	5, 2, 3
+	7, 3
×	21
v	6, 21
+	27

Note that the program for the stack machine is the Polish suffix form (Chapter 4) of the expression. The stack sequence is shown with the top of the stack on the left.

TABLE 6-1*b* THE STORE-POINTER MACHINE.

Program	Store Sequences	Pointer
3	(1, –), (2, –), (3, –), . . .	0
u	(1, 3), (2, –), (3, –), . . .	1
5	(1, 3), (2, 2), (3, –), . . .	2
+	(1, 3), (2, 2), (3, 5), . . .	3
×	(1, 3), (2, 7), (3, 5), . . .	2
v	(1, 21), (2, 7), (3, 5), . . .	1
+	(1, 21), (2, 6), (3, 5), . . .	2
	(1, 27), (2, 6), (3, 5), . . .	1

The stack is represented in memory as the contents of locations $1, 2, \ldots, k$, where k is the value of the pointer. The bottom of the stack is always at location 1. A "–" as the contents of a memory location means that the contents are possibly not known.

Table 6.1c   THE STORE-SEQUENCE MACHINE.

Program	Store Sequence
(3, 0)	(1, –), (2, –), (3, –), . . .
(u, 1)	(1, 3), (2, –), (3, –), . . .
(5, 2)	(1, 3), (2, 2), (3, –), . . .
(+, 2)	(1, 3), (2, 2), (3, 5), . . .
(×, 1)	(1, 3), (2, 7), (3, 5), . . .
(v, 1)	(1, 21), (2, 7), (3, 5), . . .
(+, 1)	(1, 21), (2, 6), (3, 5), . . .
	(1, 27), (2, 6), (3, 5), . . .

TABLE 6.1d   A SINGLE ADDRESS, SINGLE ACCUMULATOR MACHINE.

Program	Store and Accumulator	
cla 3	(1, –), (2, –), (3, –)	(A, 3)
sto 1	(1, 3), (2, –), (3, –)	(A, 3)
cla u	(1, 3), (2, 2), (3, –)	(A, 2)
sto 2	(1, 3), (2, 2), (3, –)	(A, 2)
cla 5	(1, 3), (2, 2), (3, –)	(A, 5)
sto 3	(1, 3), (2, 2), (3, 5)	(A, 5)
cla (2)	(1, 3), (2, 2), (3, 5)	(A, 2)
add (2)	(1, 3), (2, 2), (3, 5)	(A, 7)
sto 2	(1, 3), (2, 7), (3, 5)	(A, 7)
cla (1)	(1, 3), (2, 7), (3, 5)	(A, 3)
mpy (1)	(1, 3), (2, 7), (3, 5)	(A, 21)
sto 1	(1, 21), (2, 7), (3, 5)	(A, 21)
cla v	(1, 21), (2, 7), (3, 5)	(A, 6)
sto 2	(1, 21), (2, 6), (3, 5)	(A, 6)
cla (1)	(1, 21), (2, 6), (3, 5)	(A, 21)
add (1)	(1, 21), (2, 6), (3, 5)	(A, 27)
sto 1	(12, 7), (2, 6), (3, 5)	

The instructions of the store sequence are of two kinds: for 0-ary operators with address k, store the value in location $k + 1$. For binary operators with address k, take the operands from locations $k + 1$ and k and store the result in location k. Note that the instructions of this machine can be obtained directly from the store-pointer machine. The essential idea is that even though the values of the data elements may not be known at compile time, their positions in storage may be determined.

By direct translation from the store sequence machine, we obtain the program for the single address, single accumulator machine. No attempt has been made to realize even the most elementary optimizations in this code.

C. TRANSLATING EXPRESSIONS TO A STACK MACHINE

Algebraic Preliminaries

The expressions are formed in an algebra of words, where the function of each 0-ary operator is the operator itself, and the function of each binary operator is to concatenate the operands and the operator to form an expression. The carrier of this algebra is the set of valid expressions, and the algebra can be generated by the set of numerals, variable symbols, and operator symbols. Such an algebra is called a *word algebra*.

The value algebra is an algebra of values, where the function of each 0-ary operator is the numeric value of the constant or variable and the function of the binary operators is the conventional meaning of arithmetic operators. The carrier of this algebraic system is the set of natural numbers.

The evaluation of expressions is, then, a homomorphism from the word algebra of expressions to the value algebra. The idea in evaluating expressions on a stack machine is to represent the expressions as a sequence of instructions and the evaluation as a sequence of state transformations. But the sequences of instructions and state transformations are semigroups, so we cannot define homomorphisms from them to the expression and value algebras, since homomorphisms preserve operator ranks. In order to define the relationship between the program and the sequence of state transformations, we need a way of representing the algebras of expressions in the semigroup of programs and, likewise, a way of representing the algebra of values in the semigroup of state transformations. Such representations of one algebraic system in another are called *embeddings*.

We first show the embedding for the word algebra of expressions over a set of variables, X, and operators, Ω. The carrier of the semigroup that we choose for the embedding is $(X \cup \Omega)^*$, where $(X \cup \Omega)^*$ is the set of all finite sequences over $X \cup \Omega$ and the operation of the semigroup is concatenation. The mapping, f, takes each element of $X \cup \Omega$ in the expression algebra to the same element in the semigroup, and the homomorphism, ϕ, is defined by

$$\phi: x \mapsto x \quad \text{if} \quad x \text{ is in } X \cup \Omega^2 \text{ and rank } (x) = 0$$
$$\omega(\alpha_1, \ldots, \alpha_k) \mapsto \phi(\alpha_1)\phi(\alpha_2), \ldots, \phi(\alpha_u)\omega$$

It is clear that this is just the homomorphism that takes terms over a ranked alphabet to their postfix form. (Of course, the symbols in the expression algebra are understood to have an associated rank, while in the semigroup each element has rank 0, and the implicit concatenation operator has rank 2.)

Next we consider the embedding of the value algebra. The semigroup in which we embed the value algebra is the monoid of transformations of sequences of elements of the carrier of the value algebra—sequences of numbers, which we may intuitively understand as states of the stack. We add a zero element to this monoid, whose significance is that it corresponds to a (nonrecoverable) error. Each element, x, of the carrier of the value algebra is mapped to an element ρ_x of the monoid, where $\rho_x: \alpha \mapsto \alpha \cdot x$; that is, ρ_x extends the sequence α by x and does not affect the error element. Each element ω of rank >0 is mapped to an element ρ_ω of the monoid, where

$$\rho_\omega: u \mapsto a_1 \cdot \ldots \cdot a_{n-k} \cdot \omega(a_{n-k+1}, \ldots, a_n) \quad \text{if}$$
$$u = a_1 \cdot a_2 \cdot \ldots \cdot a_n, \quad n \geq k$$
$$\mapsto 0 \quad \text{otherwise}$$

The transformation ρ_ω removes the k elements from the end of the sequence corresponding to the top of the stack, $a_{n-k+1}, \ldots, a_n$, and places $\omega(a_{n-k+1}, \ldots, a_n)$ at the end of the

---

[2]That is, ϕ acts like f on elements of $X \cup \Omega$ and, since $x = f(x)$, we have elided the f in the mappings.

Figure 6-2.   The homomorphic embedding of the value algebra in the semigroup of stack transformations.

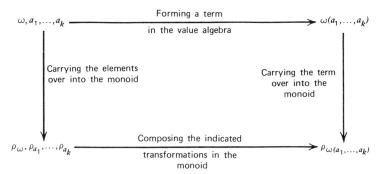

sequence. If fewer than k elements are in the sequence, the error state is entered. The error state is a zero of the transformation monoid.

We can check that this embedding is a homomorphism, since it corresponds to Figure 6-2.

Understanding the Programming Implications of the Algebra

Now let us see what the programming interpretation of this algebra is. The embedding of the expression algebra in the semigroup of the symbols of that algebra is a compilation of expressions into Polish suffix form. The embedding of the value algebra in the monoid of stack transformations (with error), under composition, takes each partial expression into a stack configuration. The mapping from expressions to values is inter-

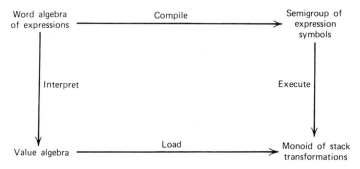

Figure 6-3.   Diagram showing the relationship compilation to interpretation.

preting the program. Finally, going directly from the semigroup of symbols to the monoid of stack transformations is an execution of the compiled program. This is shown in Figure 6-3.

If we note only that, in fact, for the subset of stack transformations corresponding to expression symbols the "load" mapping is injective, then the arrow is reversible for the submonoid, and the commutativity of this diagram expresses the correctness of the compilation.

D. TRANSLATING A STACK MACHINE TO A STORE-POINTER MACHINE

In the previous section, we showed how the embedding of the expression word algebra into a word semigroup of the symbols of that algebra under concatenation provided an instruction stream that could be understood as the code of a stack machine. We now show how the code of the stack machine can be mapped into code for a store-pointer machine. It will follow, by the composition of these mappings, that we can translate directly from the expressions to the store-pointer machine, and the correctness of the composite mapping will follow from the correctness of each of its components.

Algebraic Preliminaries

Let M be a machine with a set Q of states, a set I of inputs, and an initial state q. As described earlier (Chapters 1 and 3), each input element corresponds to a state transformation, and each input sequence corresponds to the composition of the transformations of its elements, except that we no longer require Q to be finite. Notice that this correspondence is, in fact, a homomorphism from the semigroup of input sequences to the semigroup of state transformations. We can describe this by Figure 6-4.

We may also wish to use a machine M' to simulate M. In that case, the input sequences of M will be mapped to input sequences of M', and these must carry over to appropriate correspondences between the states of M and those of M' and to appropriate correspondences between state transformations of M and those of M'. Let Q', q'_0, and I' designate, respectively, the state set, initial state, and input alphabet of M', and let ϕ' designate the homomorphism from I' to state transformations of

Figure 6-4.   The homomorphism from input sequences to state transformations.

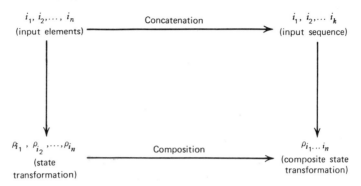

M'. We can describe the simulation of M by M' by the pair of
functions θ and h, where θ maps input sequences over I to
input sequences over I' and h maps states of M' to states of M.
We can now define a simulation of M by M'.

Definition

*A machine M can be simulated by a machine M' if there is
homomorphism, θ, from $(I)^*$ to $(I')^*$ and function $h: \bar{Q}' \underset{\text{onto}}{\to} Q$,*
where $\bar{Q}'$ is the subset of Q' reachable from q'_0 by sequences in
$(I')^*$, such that:

    (a) $h: q'_0 \mapsto q_0$
    (b) For any input sequence t in I^*, $(\phi t)q_0 = h(\phi'(\theta t)q'_0)$

Condition (a) states that the initial state of M' maps to the
initial state of M. Condition (b) states that when the input

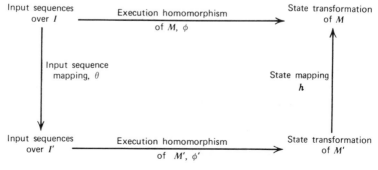

Figure 6-5.   Simulation of M by M'.

sequence θt is applied to M', θt takes M' to a state that maps to the state in M to which t leads. This simulation can be visualized as Figure 6-5.

If, in fact, the relationship between input sequences is stronger and takes sequences of length 1 over I to sequences of length 1 over I', M' clearly can simulate M.

The Relationship between the Stack Machine and the Store-pointer Machine

The stack machine has been defined as a state machine whose input sequences are the postfix forms of expressions and whose execution homomorphism takes input symbols to stack trans-formations. As a state machine, the stack machine has input set $X \cup \Omega$ and state set $V^* \cup E$, where V is the stack alphabet and E is an error state. The execution homomorphism, ϕ, is defined by

$$\phi x(v_1, v_2, \ldots, v_n) = v_1 v_2 \ldots v_n x \qquad \text{if} \quad x \text{ is in } X$$
$$\phi \omega(v_1, v_2, \ldots, v_n)$$
$$= v_1 v_2 \ldots v_{n-k} \omega(v_{n-k+1}, \ldots, v_n) \qquad \text{if} \quad \omega \text{ is in } \Omega \text{ and of rank } k$$
$$= \varepsilon \qquad \qquad \text{otherwise}$$

In words, the effect of a symbol in x, of rank 0, is to push x onto the stack. The effect of an operator of rank k is to remove the topmost k elements of the stack as its operands, operate on them, and place the result on the stack. If there are insufficient elements on the stack, an error state is entered.

The store-pointer machine consists of a store and a pointer. The store is a mapping from integers (the addresses of store locations) to values (the contents of the locations). Assign is a function that takes as input a pair (n, v), an integer, and a value, and assigns the value v to location n of the store. In other words, assign (n, v) is a mapping from S to S, where S is the set of possible stores. The input set of the store-pointer machine is $X \cap \Omega$ and the state set is $S \times (N \cup \{\varepsilon\})$, the set of store-pointer pairs. The execution homomorphism, ϕ, of the store-pointer machine is:

$$\phi x(s, n) = (s', n') \qquad \text{where} \qquad s' = \text{assign } (n, x)s \text{ and } n' = n + 1$$
$$\text{if } x \text{ is in } X$$

$$\phi\omega(s, n) = (s', n') \text{ where } s' = \begin{cases} \text{assign} \\ (n + 1 - k, \omega(s(n + 1 - k), \ldots, s(n)) \\ \text{if } \omega \text{ is in } \Omega \text{ of rank } k \text{ and } n \geq k \\ s \text{ otherwise} \end{cases}$$

$$n' = \begin{cases} n + 1 - k \text{ if } \omega \text{ is in } \Omega \text{ of rank } k \text{ and} \\ \qquad n \geq k \\ \varepsilon \qquad \text{otherwise} \end{cases}$$

Having defined both the stack machine and the store-pointer machine, we will show the mappings required to simulate the stack machine by the store-pointer machine. First, the input mapping θ is the identity function since, as we have noted, the input sequence is the same for both machines. The state mapping, h, maps the store-pointer pair, (s, n), onto a stack configuration $s(1), s(2), s(3), \ldots, s(n)$, with $s(n)$ on the top of the stack.

Now it is straightforward to show that the store-pointer machine simulates the stack machine in the sense of Figure 6-5.

E. TRANSFORMING THE STORE-POINTER MACHINE TO THE ADDRESS-PROGRAM MACHINE

Algebraic Preliminaries

Consider a cascade composition of two machines, M_1 and M_2, as shown in Figure 6-6. In this machine i is input to M_1, and the input to M_2 is the ordered pair (i, q_1), which consists of the input i to M_1 and the state q_1 of M_1.

The cascade composition can be considered as a machine M. The situation is summarized in Table 6-2.

The execution homomorphism, ϕ, of M can be described in terms of ϕ_1 and ϕ_2 by the equation $(\phi i)(q_2, q_1) = ((\phi_2(i, q_1))q_2, (\phi_1 i)q_1)$.

Alternatively, we might start with M and decompose M into the cascade decomposition. In particular, if the state set Q can

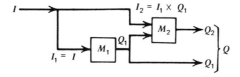

Figure 6-6.   The cascade composition of M_1 and M_2.

TABLE 6-2   INPUT AND STATE SETS OF CASCADE MACHINE.

Machine	Input Set	State Set	Execution Homomorphism
M_1	$I_1 = I$	Q_1	ϕ_1
M_2	$I_2 = I \times Q$	Q_2	ϕ_2
M	I	$Q_1 \times Q_2$	ϕ

be expressed as $Q_2 \times Q_1$ and there are functions f_1 and f_2 such that the execution homomorphism of M, ϕ, can be expressed as $(\phi i)(q_2, q_1) = ((f_2 i)(q_2, q_1), (f_1 i)q_1)$, then the execution homomorphisms ϕ_1 and ϕ_2 can be defined.[3]

Example
    Let M be the machine whose transition table is as specified in Table 6.3a.

TABLE 6-3   TRANSITION FUNCTIONS OF M AND ITS CASCADE DECOMPOSITION INTO M_1 AND M_2 (SEE FIGURE 6-6).

State	Input		
	a	b	c
P_1	P_4	P_1	P_6
P_2	P_6	P_1	P_6
P_3	P_5	P_1	P_6
P_4	P_2	P_2	P_6
P_5	P_1	P_2	P_5
P_6	P_3	P_2	P_4

(a) Transition Table for M

State	Input		
	a	b	c
q_1	q_2	q_1	q_2
q_2	q_1	q_1	q_2

(b) Transition Table for M_1

[3]For our purposes, we need not consider how to identify the decomposability of a machine. For the relevant theory, J. Hartmanis and R. E. Stearns, *Algebraic Theory of Machines*, Englewood Cliffs, N.J.: Prentice-Hall, 1965.

Suppose that we start in the initial state, P_1, of M and apply the input sequence $abcb$. Then

$$P_1 \overset{a}{\mapsto} P_4 \overset{b}{\mapsto} P_2 \overset{c}{\mapsto} P_6 \overset{b}{\mapsto} P_2$$

State	Input					
	(q_1, a)	(q_2, a)	(q_1, b)	(q_2, b)	(q_1, c)	(q_2, c)
Q_1	Q_1	Q_2	Q_1	Q_2	Q_3	Q_3
Q_2	Q_3	Q_1	Q_1	Q_2	Q_3	Q_2
Q_3	Q_2	Q_3	Q_1	Q_2	Q_3	Q_1

(c) Transition Table for M_2

The corresponding transition sequences for M_1 and M_2 are:

$$q_1 \overset{a}{\mapsto} q_2 \overset{b}{\mapsto} q_1 \overset{c}{\mapsto} q_2 \overset{b}{\mapsto} q_1$$

$$Q_1 \overset{(q_1, a)}{\longmapsto} Q_1 \overset{(q_2, b)}{\longmapsto} Q_2 \overset{(q_1, c)}{\longmapsto} Q_3 \overset{(q_2, b)}{\longmapsto} Q_2$$

The correspondence between states of M_1 and M_2 and those of M is:

$$P_1 = (q_1, Q_1) \qquad P_4 = (q_2, Q_1)$$
$$P_2 = (q_1, Q_2) \qquad P_5 = (q_2, Q_2)$$
$$P_3 = (q_1, Q_3) \qquad P_6 = (q_2, Q_3)$$

In fact, one can check that for each state P and each input i where $P \overset{i}{\mapsto} P'$, then P corresponds to (q, Q), $q \overset{i}{\mapsto} q'$, $Q \overset{(q, i)}{\longmapsto} Q'$ and P' corresponds to (q', Q').   #

The decomposition of M, when it exists, is useful, because it allows us to run the input sequence through M_1, computing the input sequence for M_2 as we do this. Then, in a second pass, we run the computed input sequence for M_2 through M_2. The programming interpretation of this two-stage computation is that a computation using M_1 will develop the storage assignments to simulate the stack machine, and the execution of the code with the known storage assignments can be done in M_2. Stated otherwise, M_1 acts as a compiler, taking as its source code a sequence in I^* and generating code with fixed storage locations in $(I \times Q_1)^*$ for M_2, and M_2 runs this object program.

The fact that the cascade decomposition can be applied by first doing all computation on M_1 and then doing all computation on M_2 is contained in the following proposition.

Proposition. Suppose that M can be decomposed into a cascade decomposition using M_1 and M_2, as in Figure 6-6, so that $(\phi i)(q_2, q_1) = ((\phi_2(i, q_1))q_2, (\phi_1 i)q_1)$. Then there is a homomorphism, τ, mapping input sequences over I into transformations of sequence-state pairs, where the sequences are over $I \times Q$ and the state is in Q_1. The homomorphism is defined by:

$$\tau: i(t_2, q_1) \mapsto (t_2(i, q_1), \phi_1 i q_1)$$

and the execution homomorphism ϕ for an input sequence t applied to state (q_2, q_1) is given by

$$(\phi t): (q_2, q_1) \mapsto (\phi_2 t_2 q_2, q_1')$$

where $(t_2, q_1') = \tau t(\lambda, q_1)$ and λ is the null sequence.

INFORMAL EXPLANATION.   If M has a cascade decomposition so that each state q of M is represented by the state pair (q_2, q_1), the effect ϕ of each input i can be described as the effect ϕ_1 of i applied to M_1 in q_1 and the effect ϕ_2 of (i, q_1) applied to M_2 in q_2. Furthermore, a sequence t of inputs to M can be homomorphically mapped by τ to compute a sequence t_2 so that the effect applying t to M_1 and then applying t_2 to M_2 is the same as the effect of applying t to M.

Proof.   The proof is by induction on the length of input sequences.

BASIS. $t = i$, a sequence of length 1.

$$\tau i(\lambda, q_1) = ((i, q_1), \phi_1 i q_1)$$

by definition of τ, and thus $t_2 = (i, q_1)$. Applying i to M_1 and t_2 to M_2 yields

$$(\phi_2(i q_1) q_2, \phi_1 i q_1)$$

and, by the hypothesis of the proposition, this is

$$\phi i(q_2, q_1)$$

So, for sequences of length 1, we may apply t to M_1 and t_2 as computed in (the first component of) τt to M_2.

Induction. Assume that $\tau t(\lambda, q_1) = (t_2, q_1')$ and that $\tau(\tau \cdot i)(\lambda, q_1) = (t_2^*, q_1'^*)$. We must show that $\phi(t \cdot i)(q_2, q_1) = (\phi_2 t_2^* q_2, q_1'^*)$. We have $t_2^* = t_2 \cdot (i, q_1')$ and $q_1'^* = \phi_1 i q_1'$ directly from the definition τ given in the statement of the proposition. We now proceed to evaluate the effect of $t \cdot i$ on M, $\phi(t \cdot i)(q_2, q_1)$.

$$\phi(t \cdot i)(q_2, q_1) = \phi i(\phi t(q_2, q_1)) \tag{6-1}$$

$$= \phi i(\phi_2 t_2 q_2, q_1') \tag{6-2}$$

$$= (\phi_2(i, q_1')(\phi_2 t_2 q_2), \phi_1 i q_1') \tag{6-3}$$

$$= (\phi_2(t_2(i, q_1')) q_2, \phi_1 i q_1') \tag{6-4}$$

$$= (\phi_2 t_2^* q_2, q_1'^*) \tag{6-5}$$

Explanation of Equations 6-1 to 6-5.

    (a) Applying $t \cdot i$ is the same as applying t and then applying i. Stated otherwise, $\phi(t \cdot i) = \phi(t) \cdot \phi(i)$; that is, the effect of execution, ϕ, is a homomorphism.

    (b) This is the inductive assumption.

    (c) This is the hypothesis of the theorem, that M has a cascade decomposition. Note that this is used only for an input sequence of length 1 (i.e., an input symbol).

    (d) This is just the execution homomorphism $\phi_2(t_2 \cdot (i, q_1')) = \phi_2(t_2) \cdot \phi_2(i, q_1')$.

    (e) This is just the definition of t_2^* and $q_1'^*$.

Applying the Cascade Decomposition and Translation Homomorphisms to Replacing the Store-pointer Machine by the Address-program Machine

In the cascade decomposition of the store-pointer machine, the value of the pointer is computed at compile time and is included in the address of the instruction. We show that the store-pointer machine has the appropriate form for such a cascade decomposition.

    Recall that the execution homomorphism of the store-pointer machine is given by:

$$\phi x(s, n) = (s', n') \qquad \text{where} \qquad s' = \text{assign } (n, x)s \text{ and } n' = n + 1$$
$$\text{if } x \text{ is in } X$$

and

$$\phi\omega(s, n) = (s', n') \text{ where } s' = \begin{cases} \text{assign} \\ (n+1-k, \omega(s(n+1-k), \ldots, s(n)))s \\ \text{if } \omega \text{ is in } \Omega \text{ of rank } k \text{ and } n \geq k \\ s \text{ otherwise} \end{cases}$$

$$n' = \begin{cases} n+1-k & \text{if } \omega \text{ is in } \Omega \text{ of rank } k \text{ and} \\ & n \geq k \\ \varepsilon & \text{otherwise} \end{cases}$$

This can be seen to be of the form:

$$\phi i(s, n) = (f_2 i(s, n), f_1 in)$$

where

$$f_1 in = \begin{cases} n+1 & \text{if} & i \text{ is in } X \\ n+1-k & \text{if} & i \text{ is in } \Omega \text{ of rank } k \text{ and } n \geq k \\ \varepsilon & \text{otherwise} \end{cases}$$

and

$$f_2 i(s, n) = \begin{cases} \text{assign } (n, i) \text{ if } i \text{ is in } X \\ \text{assign } (n+1-k, i(s(n+1-k), \ldots, s(n)))s \\ \quad \text{if } i \text{ is in } \Omega \text{ of rank } k \text{ and } n \geq k \\ s \quad \text{otherwise} \end{cases}$$

F. TRANSLATING THE STORE-SEQUENCE MACHINE TO THE SINGLE-ADDRESS, SINGLE-ACCUMULATOR MACHINE

In this translation, each step of the store-sequence machine is directly replaced by a sequence of instructions of the single-accumulator machine.

Instructions of the form (v, n) are replaced by sequences:

CLA v

STO $n+1$

Instructions of the form (θ, n) are replaced by sequences:

CLA n

OPθ $n+1$

STO n

where OPθk takes its first operand from the accumulator and its second operand from location k and leaves the result in the accumulator.

The replacement of the store sequence machine by the single-address, single-accumulator machine is another example of simulation of one machine by another, as discussed in Section D. Here the input mapping θ takes each input to the store-sequence machine and expands it to a sequence of inputs to the single-accumulator machine. After the sequence of inputs is applied, the contents of the store of the single-address, single-accumulator machine are the same as the contents of the store of the store-sequence machine.

REVIEW

We considered some detailed examples that illustrate the concepts and techniques of discrete structures in computer science. The student is now ready to read directly much of the technical literature of the subject; an appropriate starting place is the original Burstall and Landin paper on which most of this chapter is based.

PROBLEMS

P 1. Let G be the class of labeled digraphs, as defined in Chapter 2.

(a) Develop a coding mapping, C, from G to the class of (labeled, directed, ordered) trees, T. Is the mapping C injective?

(b) Give a representation (partial) function, R, from T to C.

(C) Show that $C \circ R$ is the identity on G.

(d) What can you say about $R \circ C$?

P 2. Using the results of Problem 1 and the coding given in Section A, develop a coding from graphs to binary trees.

E 3. (a) Using the grammar of arithmetic expressions, develop the derivation tree of $(6 + v) \times (4 + v)$.

(b) Using the semantic rules, show the evaluation of this tree for $v = 5$.

(c) Show the stack machine program and sequence of stack states in evaluating the expression of part a.

(d) Continue this example and develop the program and store sequences for the store-pointer machine, the store-sequence machine, and a single-address, single-accumulator machine.

SUGGESTED PROGRAMMING EXERCISES

1. Using arrays as the only data structure, develop a program to compute the prefix form representation of a tree, where the tree itself is coded as an array.

2. (a) Using arrays as the only data structure, develop a program to compute the infix representation of a tree.
   (b) Compare your program of part a with a recursive program that has trees as data structures.

3. Write a program to translate arithmetic expressions into a program for a single-address, single-accumulator machine.

INDEX

Page numbers in italic type represent either the main definition or principle entry of the item.